# A UNIFIED THEORY OF ADVERSARIAL DYNAMICS

30 SHARED PRINCIPLES OF STRUGGLE DRAWN
FROM DOZENS OF DISTINCT FIELDS TO HELP
YOU WIN IN ANY ARENA

## CHRIS GRIGGS

**GREGOROS**

*A Unified Theory of Adversarial Dynamics*

*30 Shared Principles of Struggle Drawn from Dozens of Distinct Fields to Help You Win in Any Arena*

Published in the United States by Gregoros LLC, Tallahassee, Florida

ISBN 979-8-9997336-0-3 (hardback)

ISBN 979-8-9997336-1-0 (paperback)

ISBN 979-8-9997336-2-7 (ebook)

Cover and interior design by Chris Griggs

First printing, 2025

# CONTENTS

*For those who fight — with purpose, compassion, and resilience — for good.*

# CODE IN THE CONFLICT
## INTRODUCTION

*"War has its own grammar."* — Carl von Clausewitz

A ll domains in which there is struggle between two opposing and adaptive forces—from war to cybersecurity to medicine to the animal kingdom—share an underlying set of patterns that is *remarkably* consistent.

A few years ago, I had an epiphany of sorts. I was attending an evening hapkido class at a local martial arts studio. On this particular night, we had taken a break from our usual technique drills. The instructor had all the students sit down, and he spent some time educating us about the foundational principles of hapkido.

All martial arts have principles that serve as the unifying philosophy determining which techniques belong and why. This was my first exposure to the idea of *hard* versus *soft* styles. Hard arts such as karate and taekwondo focus on striking—punches and kicks—while soft arts such as jujutsu and hapkido focus on grappling, joint locks, or throws. Soft styles emerged as tactical adaptations for unarmed fighters who had to subdue armored opponents—samurai—against whom striking was ineffective. In like manner, functional realities shaped every other art.

In hapkido, three interwoven principles guide technique:

- *Harmony* blends with an opponent's movement, using their own momentum against them. When someone pushes against us, the instinct is to push back. Harmony says that if they push, you *pull*. And because the opponent was expecting resistance, the act of pulling catches them off-guard and destabilizes them, allowing you to seize the initiative.
- *Circular motion* prefers arcs over lines, deflecting force and opening new angles of control. Stopping a punch coming at your face with direct opposing motion requires at least the same amount of force as the punch itself. But deflecting that same punch away from its target with a simple angled tap requires almost no force at all.
- *Flow* demands continuous adaptability; like water, the practitioner shifts to fit the container, overcoming obstacles by going around them or avoiding harm by continuous movement. In modern vernacular, we might think of this as "staying fluid" during a fight.

I absorbed everything the instructor said about the principles with great interest. The class ended, and I went home. I'm not sure if what happened next was coincidence, fate, or just pure old-fashioned luck.

I usually like to spend a little time reading before going to bed. On this particular night, I picked up my well-worn copy of Dale Carnegie's classic *How to Win Friends and Influence People*. There is a scenario in the book that describes how a commercial truck salesman would win over customers that were initially skeptical and antagonistic. Instead of defending his product when someone criticized it, he'd agree—even compliment the competitor's product. If a customer said, "Your truck is no good—I'm loyal to the other brand," the salesman would respond, in effect, "I actually admire your loyalty. I can see why you like those trucks. Here's what I like about them, too

..." This caught the customer off-guard, disarmed the tension, and shifted the conversation from confrontation to curiosity. With resistance lowered, he could then calmly highlight the strengths of his own trucks. Rather than winning a debate, he built rapport—and sales followed.

Having returned home from the hapkido class maybe only an hour before, the epiphany hit: *Dale Carnegie's truck salesman was using hapkido principles.*

Sure, he wasn't using the principles physically. But he was absolutely using them psychologically.

- He employed the *harmony* principle by avoiding direct disagreement with the customer's opinion and instead aligned with their momentum.
- He employed the circular motion principle by deflecting and opening new avenues of influence.
- He employed the flow principle by staying adaptive, engaged, and emotionally attuned to build personal rapport with the customer.

The result was a win. It was all there. The truck salesman's approach was a direct cognitive corollary to the principles of hapkido.

Thus was born for me in that moment—however raw and nascent—the idea of a Unified Theory of Adversarial Dynamics (UTAD). I became fascinated with the idea that forces at play in physical conflict were actually only one manifestation of a deeper meta-truth that could also manifest in other scenarios where opposing forces clashed.

∾

AFTER SEVERAL YEARS OF RESEARCH AND REFLECTION, I CAN confidently say there is indeed an underlying code in conflict. There is a structure to struggle. Making a connection between very similar domains such as war and martial arts is fairly intuitive and doesn't require a large mental leap. These both represent conceptually similar physical conflict manifestations, though clearly on different scales. But seeing connections between, say, *red-team cyber operations* and *the stalking tactics of Nile crocodiles in capturing large prey*? Not quite as intuitive. This type of connection eludes us despite the fact that struggle patterns are all around us. You can feel them in the ebb and flow of a chess match between grandmasters, where each move radiates tension and possibility. You can hear them in the rhythm of a courtroom cross-examination, where questions are weapons and silence is a trap. You can witness them in nature in the coordinated attack of a wolf pack. In the relentless volleys of a political debate. In the digital pulse of a cyber breach. In the determined words of a cancer patient who tells their doctor that they want to fight the terrible disease on all fronts.

These contests are indeed diverse—spanning war, sport, business, biology, and beyond—but beneath the surface lies something consistent. The same underlying dynamics emerge again and again in any environment where outcomes are contested, where intention meets resistance, and where two or more agents seek to impose incompatible wills on one another. We call these phenomena *adversarial dynamics*. Whether the adversary is another person, another team, a system, an animal, or even invasive plants (yes, *plants*), the underlying structure of strategic engagement remains surprisingly consistent.

UTAD is the culmination of deep research into over two dozen distinct domains. The research entailed extensive cross-domain pattern analysis and synthesis. It draws on centuries of data and wisdom where we see the purest presentations of adversarial dynamics. Across so many domains we would expect to see hundreds—maybe *thousands*—of governing principles. *We see only thirty.* Again and again. No matter the fighters. No matter the arena. *Thirty.*

I do not argue that all conflict is the same. Domains differ. You will see that not all thirty principles apply to every manifestation of conflict. There are nuances that matter. Sports coaches might like to use war analogies, but we all understand that sports competition has obvious and important differences from armed conflict. However, there are parallels and principles that are generally applicable. And I argue that the underlying dynamics—the "physics of contest"—can be mapped. UTAD is that map.

∿

THIS BOOK EXPLORES EACH OF THE THIRTY PRINCIPLES. THE GOAL IS NOT to overload you with theory but to equip you with transferable wisdom: conceptual clarity, real-world examples, and applied tools. Along the way, you'll encounter historical military campaigns, sports collapses, courtroom tactics, hacker exploits, animal behavior, and psychological gamesmanship. Every example is selected to reinforce a principle that transcends its specific context.

We draw from Sun Tzu and Clausewitz. We draw from John Boyd, John Keegan, and Machiavelli. But we also draw from viruses and NBA offenses. From cutting-edge artificial intelligence warfare and the social maneuvering of elephants. Because once you see the structure of adversarial dynamics, you begin to realize that everyone is playing the same game with different pieces.

The idea that there are patterns to conflict that transcend domains is not new. What I claim as novel within UTAD is the effort to systematically extract those principles in their entirety from such a large sampling of domains and then cross-apply them with rigor.

Most strategy writing suffers from tunnel vision: it teaches you how to negotiate, or how to win at chess, or how to lead a company, or how to fight in the ring. But it rarely crosses boundaries. It rarely asks: What can military strategists learn from *the field of pathology* to more effectively combat an insurgent force? How can Chief Information Security Officers (CISOs) harness lessons from *Brazilian Jiu-Jitsu* to better secure their networks? When will trial attorneys recognize the value of studying *alpha conflict dynamics in populations of baboons* to win bigger for their clients? (With deepest apologies to my lawyer friends).

But better yet, why don't the individuals above go even broader?

Why don't they study adversarial dynamics across *dozens* of domains? They should. And now they can. That's what UTAD is. It offers readers an opportunity to look to outside domains where they may find inspiration for novel or improved approaches to struggle within their own.

~

We have also entered a time in history in which old domain distinctions are breaking down anyway. Business leaders are now targets of political violence. Militaries operate in digital space. Journalists wage narrative warfare. Politicians hire brand strategists. The lines between domains are continually blurring, and success now depends less on mastery within a narrow field and more on the ability to recognize patterns of power and resistance across all of them.

If you speak one language, you understand the mechanics of that particular language. If you speak two languages, you no doubt see some linguistic structural parallels between them. But if you speak dozens of languages, you may better understand the underpinnings of something that Noam Chomsky called Universal Grammar. You know the language of language that Chomsky argued is hardwired into all humans as a language operating system, of sorts. My Arabic instructor in college spoke five or six languages, and he was actively learning two more at the time. I recall him mentioning that each new language was easier to acquire because, at some point, he recognized that language is collectively formulaic. And understanding that universal formula helped him sharpen and refine each individual language he spoke.

What if UTAD could do the same? What if studying across adversarial domains conveys something like a *Universal Conflict Grammar*? A strategic literacy of fighting and winning? What if that literacy enabled you to sharpen your own actions within the adversarial domains in which you operate?

~

THIS BOOK IS DESIGNED TO GIVE YOU SUCH A LITERACY. WHAT DOES such literacy do for you?

Let's start with what it *won't* do. UTAD won't tell you whether you should go for a full house or a flush with your poker hand. It won't tell you how to secure your company's computer network in light of a new ransomware threat targeting your industry. It won't tell you which play to call when the youth football team you're coaching is third-and-nine on your own five-yard line and you have to play a lousy quarterback because he's the son of your team sponsor. These are tactical decisions.

So what *does* UTAD do? Four things:

1. **UTAD educates.** It provides strategic literacy—making visible the forces of pressure, initiative, exposure, tempo, deception, and collapse, and turning knowledge into actionable fluency.
2. **UTAD translates.** It offers a common language that allows for better communication and understanding regarding concepts and phases of conflict.
3. **UTAD orchestrates.** It supplies a map for planning contested engagements. In doing so, it helps unify actions across teams that are operationally heterogeneous yet aligned around a shared strategic purpose.
4. **UTAD evaluates.** It serves as a diagnostic checklist, auditing campaigns both during and after a conflict.

Together, these uses form a lattice: *education, translation, orchestration,* and *evaluation.* Each is distinct, but all serve the same larger purpose—equipping individuals and systems to operate more intelligently within dynamic, resistant, and often hostile environments. UTAD is not merely a framework for understanding conflict. It is a framework for *engaging it well.*

And engaging well matters now more than ever. In a post–Information Age defined by the emergence of artificial intelligence, where the ability to access and analyze large amounts of information is no longer a decisive advantage, victory depends on *operational outmaneuvering*: acting faster, more effectively, and more strategically than your adversaries. To the degree that UTAD enhances your ability to execute with precision, anticipate moves, and adapt swiftly, it becomes not just a tool, but a key differentiator. Today's competitor must not only analyze the battlefield— they must outthink and outmaneuver with flawless execution.

Whether you're a strategist or a team leader, a founder or a field commander, an analyst or a coach—UTAD will sharpen your lens. It will help you see patterns sooner, make moves cleaner, and recover faster.

In short, it will help you win in your own arena.

## Sources

- Carnegie, Dale. 1936. *How to Win Friends and Influence People*. New York: Simon & Schuster.
- Chomsky, Noam. 1975. *Reflections on Language*. New York: Pantheon Books.
- Clausewitz, Carl von. 1984. *On War*. Translated by Michael Howard and Peter Paret. Princeton, NJ: Princeton University Press.
- Pellegrini, John. 1999. *The Official Combat Hapkido Manual*. Boca Raton, FL: International Combat Hapkido Federation.
- Pellegrini, John. 2000. *Combat Hapkido: The Martial Art for the Modern Warrior*. Boca Raton, FL: National Self-Defense Institute.

# BUILDING A CREDIBLE THEORY OF STRUGGLE

## UTAD'S EVIDENCE-BASED FOUNDATION

*"A wise man, therefore, proportions his belief to the evidence."* —
David Hume

I s there a shared underlying structure to all forms of conflict? A hidden code that repeats across vastly different domains? Following my epiphany around the similarities between hapkido and sales, I suspected there was. The work that followed was an attempt to test that hypothesis by examining a wide range of domains—warfare, law, medicine, cybersecurity, sports, artificial intelligence, and others—to see whether patterns would hold. What emerged was a recurring structural logic: certain principles appeared again and again, despite the contextual differences. The *result* of that process was a structured framework, but the deeper insight was theoretical. UTAD proposes that conflict, in all its complexity, is governed (or at least *exemplified*) by a finite set of dynamic patterns that recur regardless of the specifics of form. That proposition—the idea that there is a deep, transferable architecture underlying adversarial systems—led me to call this a *unified theory* instead of a *framework*. Not a theory in the sense of a scientific theory that predicts outcomes through mathematical formalism, but in the classical tradition: a conceptual model that reveals structure across phenomena.

I also spent time weighing the term *unified* against *universal*. At first glance, "universal" might seem to promise more—an appeal to absolutes. But I came to view it as the wrong word. Not every principle in this book appears in every domain. A computer virus has no team morale to protect; an ant colony does not create battle contingency plans the way humans do. The point is not that each principle is omnipresent, but that the set as a whole arises consistently across systems with meaningful regularity. "Unified" captures the more precise reality: that while domains express conflict in domain-specific ways, the foundational logic behind their dynamics can be reconciled into a shared structure. The principles in this book don't claim universality. They represent a *unified field* of adversarial behavior drawn from empirical comparison. It's that convergence, not any one example, that gives UTAD its explanatory power.

Similarly, I considered whether to use the terms *patterns* or *strategies* to describe the thirty dynamics in this book. *Patterns* seemed too incidental, since many of these dynamics are employed with clear intentionality by thinking actors. *Strategies* seemed too deliberate, since some dynamics also appear in life forms and systems that do not reason as humans do. *Principles* strikes the right balance: they are fundamental dynamics that recur across adversarial systems—sometimes chosen deliberately, sometimes expressed unconsciously, but always shaping the contest.

### Foundations and Lineage

No theory emerges in isolation. UTAD is built atop a foundation laid by strategists, systems thinkers, psychologists, and theorists who each captured fragments of what conflict truly is. Their models described how pressure accumulates, how resistance behaves, and how outcomes unfold under constraint. UTAD would not exist without their work.

Clausewitz described war as the collision of force and will. Boyd injected tempo and adaptation into adversarial thinking through the OODA loop. Chris Argyris exposed how organizations resist learning through defensive routines. Daniel Kahneman mapped how human cognition buckles under bias, inertia, and bounded rationality. Lawrence Freedman reframed strategy as an emergent narrative shaped by pressure. Martin van Creveld showed how command and control systems break under friction. Each of these thinkers illuminated a core dynamic of engagement, but always within the boundaries of a particular domain.

UTAD does not discard those insights. It organizes them. Where others explain conflict *in* context, UTAD seeks what endures *across* context. It is not meant to replace these traditions. It is meant to *bridge them structurally*. It honors what came before not by imitating it, but by building scaffolding around the most enduring insights—so they can be tested, refined, and transferred.

$\sim$

THE FOLLOWING EXPLANATIONS OF THE APPROACH TO DEVELOPING UTAD may be criticized as dry or overly academic for a general readership. But I felt them important to include to convey a sense of the breadth and depth of the underlying research, to help show the validity of the theory. If you wish to skip ahead, that is okay. But I don't recommend it. Understanding the design will give you a greater appreciation for the gauntlet of analysis each principle had to undergo to make it into these pages.

## Defining Adversarial Dynamics

Not all resistance qualifies as adversarial. A rock placed in a stream obstructs flow. Over time, it may even change shape—eroded, rounded, or cracked by pressure. But these changes do not occur in active response to the stream. The rock does not adjust itself, shift posture, or restructure its resistance based on what confronts it. It passively absorbs force. It neither escalates nor adapts. There is no exchange—only collision.

Adversarial dynamics require more than collision. They require interaction between *adaptive* forces—systems that not only persist under pressure but respond to it. That response may take many forms: strategic adjustment, behavioral reconfiguration, chemical suppression, procedural redirection, or mutation. It may be deliberate or emergent, conscious or unconscious. What matters is not an adversary's level of awareness, but its structural response. A force becomes adversarial when it *behaves*.

Both sides must meet this threshold. If only one force adapts while the other remains inert, degrading, or unresponsive, the interaction may be dangerous—but it is not adversarial. It is not engagement.

This standard holds across all UTAD conflict pairings. Even in intrapersonal conflict, one behavioral pattern interrupts or suppresses another. The conflict manifests through avoidance, compulsions, or recursive sabotage. These internal dynamics are adversarial because they interfere behaviorally.

In algorithmic competition, adversarial loops train against one another by reshaping error and prediction. In microbial or ecological competition, suppression occurs through chemical exclusion, resource monopolization, or spatial dominance. Neither of these requires cognition. But both of them qualify.

⁓

UTAD THEREFORE DEFINES ADVERSARIAL DYNAMICS AS FOLLOWS:

*Adversarial dynamics occur when two or more adaptive forces oppose each other.*

This definition is compact, but each word is deliberate:

- **"Adversarial dynamics"** is the term being defined.

- **"Two or more"** reflects UTAD's relational minimum. Conflict must occur between distinct behavioral configurations, even if those configurations exist within the same mind or system.
- **"Adaptive"** is the critical filter. It excludes passive blockage, inert form, and background friction. An adaptive force is one that responds. It modifies its structure, behavior, or persistence in reaction to pressure. It need not be intelligent, but it must behave.
- **"Forces"** is used instead of "agents" to reflect UTAD's inclusivity. Not all adversarial participants are conscious or bounded. Forces may include human actors, algorithms, immune systems, institutions, plant systems, or psychological mechanisms. What unifies them is not form, but function. This is distinct from most other definitions of adversarial dynamics, which specify the need for *intelligent* opponents.
- **"Oppose each other"** completes the definition. The relationship must be structured tension—resistance that runs counter to another force's persistence or trajectory. This does not require zero-sum conditions. It requires that each side make the other's success more difficult.

**Multi-Domain Foundations**

With adversarial dynamics properly defined, I began to identify as many distinct domains as possible in which adversarial dynamics clearly appear. What emerged was a surprisingly robust sample of adversarial life across scale, medium, and form. About two dozen domains were selected for focused analysis. The result was a collection that spans:

- **Armed Conflict** (with examples spanning western and eastern philosophies, various scales of war, as well as symmetrical and asymmetrical conflict)

- **Business Competition**
- **Martial Arts** (including real-world self-defense as well as sport forms, to include Krav Maga, Karate, Brazilian Jiu-Jitsu, Hapkido, Wrestling, Boxing, Mixed Martial Arts, Muay Thai, Judo, Kali/Eskrima/Arnis, Combat Sambo, Jeet Kune Do, Systema, Modern Army Combatives, Taekwondo, European Broadsword, and Olympic Fencing)
- **Institutional and Political Resistance** (where individuals or small groups have to fight against an entrenched bureaucracy)
- **Law Enforcement** (including strategic operations and tactical applications)
- **Sports** (*excluding* fighting sports, but including American football, basketball, baseball, ice hockey, soccer, tennis, Formula 1 Racing, volleyball, rugby, and cricket)
- **Games** (including chess, Go, poker, pool, StarCraft, backgammon, Magic: The Gathering, and others)
- **Cybersecurity** (spanning a broad range of configurations to include Blue Team vs. Red Team and real-world threats from script kiddies to Advanced Persistent Threats and nation-state actors)
- **Political Campaigns**
- **Internal Conflict and Subconscious Sabotage** (e.g., intrapersonal dynamics)
- **Medicine** (including both humans fighting pathologies as well as immune systems fighting pathologies, i.e. immunology)
- **Predator-Prey in Nature**
- **Invasive Plants** (to include human containment of them as well as how invasive plants compete against native flora)
- **Humans vs. Animals** (e.g., hunting, pest control, zoonotic threat)
- **Artificial Conflict Systems** (e.g., AI vs. AI, humans vs. AI)

How we quantify the *exact* number of domains studied greatly depends on subjective judgment over what constitutes the boundaries of a domain. In the martial arts example above, we have seventeen martial arts. Ultimately, I counted martial arts as only two domains, accounting for applications for sport as well as self-defense. Beyond that distinction, the differences in principles become negligible. Conversely, an area of study like "humans vs. animals" cleanly breaks down into a minimum of three separate domains, to account for hunting, pest control, and invasive species control. In these cases, the structural differences are substantial enough to warrant independent evaluation as distinct domains.

This diversity in domain selection is not academic flourish. It is methodological foundation. If conflict truly follows recognizable patterns, then those patterns should sustain even across such diversity.

## Overview of the UTAD Axes

Having identified the domains, the next task was to figure out the meaningful differences between them. To analyze adversarial dynamics across domains, I developed six structural axes. These axes define several formal dimensions along which an engagement can vary. My hope was that the selected domains would provide saturation across the axes. I was not disappointed in the results.

The six axes are as follows:

1. **Agent Type** – What is the fundamental nature of each opposing force?
    - Human — Capable of abstract reasoning, symbolic modeling, and strategic planning.
    - Animal — Behaves primarily through instinct, conditioning, and biologically encoded pattern recognition.
    - Artificial Intelligence — Adapts algorithmically; capable of optimizing, simulating, or competing through synthetic rules.

- ○ Unconscious Life — Exhibits behavioral persistence without cognition; includes viruses, cancer, fungi, and plant systems.
2. **Adversarial Configuration** – How is each force internally organized?
    - ○ Single Agent — One cohesive actor, system, or process.
    - ○ Coordinated Group — Multiple aligned actors behaving under shared structure or command.
    - ○ Distributed Network — A loosely coupled system whose behavior emerges from collective interaction.
    - ○ Fragmented System — A disjointed, incoherent force whose resistance stems from procedural inertia, entropy, or internal contradiction.
3. **Pairing Type** – Which forces oppose each other? Note: not all pairings are equally weighted. Some pairings are critical to UTAD (e.g., human vs. human) while other pairings (e.g., AI vs. animal) are at present not as widely observed. But they are included here for formal comprehensiveness:
    - ○ Human vs. Human
    - ○ Human vs. Animal
    - ○ Human vs. AI
    - ○ Human vs. Unconscious Life
    - ○ Human vs. Self
    - ○ AI vs. AI
    - ○ AI vs. Animal
    - ○ AI vs. Unconscious Life
    - ○ Animal vs. Animal
    - ○ Unconscious Life vs. Unconscious Life
4. **Strategic Posture** – What role does each force adopt within the engagement?
    - ○ Offensive — Seeks to dominate, disrupt, or impose.
    - ○ Defensive — Seeks to preserve, contain, or absorb pressure.

- ○ Hybrid — Combines offensive and defensive behaviors, either simultaneously or adaptively. Most engagements fall into this category.
- ○ Passive-Obstructive — Does not seek victory directly. Resists through delay, friction, or misdirection.

5. **Constraint Environment** – What governs or limits the behavior of each force? Do legal, moral, or procedural constraints shape how the engagement unfolds? Note that these can be—and often are—asymmetrical, meaning one "side" is playing by different rules than the other side. Law enforcement vs. criminals is a great example of an asymmetrical constraint environment.
   - ○ No Constraint — The force operates without imposed limits on action.
   - ○ Ethical or Moral Constraint — The force is bound by internalized values, norms, or beliefs.
   - ○ Legal or Rules-Based Constraint — The force is bound by external codes, laws, or procedural obligations.

6. **Victory Condition** – What outcome does each force seek? This is another axis that is often asymmetrical.
   - ○ Elimination — Destroy or remove the opposing force.
   - ○ Control — Constrain or dominate the opponent without destroying it.
   - ○ Survival — Persist under pressure without defeat.
   - ○ Escape — Disengage successfully from the engagement.
   - ○ Resolution — End the adversarial state through integration, reconciliation, or repair.
   - ○ Decision — Win through a structured system of judgment (law, sport, vote).
   - ○ Optimization — Use the adversary to improve one's own system or capabilities.

Together, these axes form the skeleton of UTAD's comparative logic—the scaffolding that ensures UTAD is maximally applicable

to conflict in all its manifestations. The purpose of these axes is not
to flatten difference, but to locate recurring structure where the
surface features differ. Are there more axes that *could* have been
included? Probably. But these felt like the main ones needed to
ensure the domains were representative of a variety of configurations of conflict:

- All relevant pairing types are either directly represented or
  plausibly modeled.
- Each posture category is exemplified, often in asymmetric
  pairings.
- Constraint environments vary not just across domains, but
  within engagements themselves.
- Victory conditions span the full spectrum: survival,
  elimination, optimization, decision, and more.

## Emerging Conflict Forms

Strategists often look to the ancient wisdom of Sun Tzu for insights
into conflict. This is perfectly understandable. In my research into
every principle in this book, Sun Tzu's insights surfaced time and
again. They are truly timeless and often domain-generalizable themselves. However, I'm excited by the prospect of the latest wave of technology and what it could mean for our understanding of conflict.

In particular, AI vs. AI configurations provide some of the purest
adversarial dynamics ever witnessed. What transpires in human
warfare is often slow, and clear causal mechanisms are elusive
because of interfering variables that are not directly related to the
conflict. Conversely, AIs do battle in relatively controlled environments. Intense conflict with hundreds of moves and countermoves
can transpire in seconds or minutes. Perhaps my favorite domain to
study as part of this effort was Generative Adversarial Networks
(GANs).

AI vs. Animals is another emerging conflict form that offers inter-
esting dynamics. Recently, researchers at the University of Florida

developed AI-enabled robot rabbits to help combat invasive pythons in the Everglades.

These new types of struggle are promising in the insights they can provide. I was very pleased to include a few AI-centric domains in this study to account for emerging types of conflict.

## Going Forward

This chapter was meant to provide some insights into how UTAD was structured and researched. My hope is that, as you move on to reading about the principles themselves, you do so with an appreciation for the depth of evidentiary support underlying each one. If nothing else, UTAD is validation that these are—in fact—transcendent principles that can easily cross domains.

The principles are organized *somewhat* sequentially along the lifecycle of an engagement: from preengagement dynamics and entry conditions, to escalation and active conflict, to resolution and postengagement restructuring. Some principles exist across the entirety of the temporal spectrum and run in parallel to each other. In the case of real conflict, the dynamics can also cause the principle implementation to repeat, or to jump several steps backwards or forwards on the timeline. It's impossible to put them in a perfect order. Believe me, I tried. They have been arranged and re-arranged and then second- and third-guessed. Ultimately, the nuance of order as presented here is of little consequence if you understand the essence of the principles.

I similarly wrestled with defining the boundaries of each principle, as there is extensive overlap or interplay between many of them. For example, there's an argument to be made that *Shape the Environment to Your Advantage* (Principle 11) and *Exploit the Environment* (Principle 15) could be consolidated into a single principle. They both involve the contextual environment (e.g., physical, regulatory, etc.) in which conflict unfolds. But one is executed preengagement and the other is executed during an active engagement. *Designing* a fighter jet and *flying* a fighter jet both involve jets, but they require fundamentally different skills and approaches. These types of decisions could have resulted in a reduction or expansion of the number of principles. But the same underlying elements would have to be articulated and unpacked, regardless of configuration. As with the *order* of principles, the *number* of principles and their boundaries matter less than the collective whole of the lessons they contain.

UTAD's thirty principles are also not intended to map one-to-one onto every named strategy or doctrine. Many real-world strategies are *composites* — bundles of principles applied together. For example, *maneuver warfare* in the military combines timing, adaptability, stretching defenses, and indirect attack. UTAD works at a more fundamental level of abstraction: it identifies the elemental dynamics that underlie composite doctrines.

Each principle chapter follows a specified pattern. First, the principle and its subcomponents are defined and explained. Next, a *positive* real-world example is presented, where the principle was implemented with success. After that, a negative real-world example is shared, where a failure to apply the principle (either entirely, correctly, or consistently) led to a bad outcome. I've made an effort to include examples from a variety of domains across the UTAD spectrum. Finally, some practical questions are provided to help you think through implementation of each principle in your own domain.

The thirty principles that follow are not offered as absolute laws. But they *are* recurrent. Many can be applied in strategic contexts (e.g., an entire war campaign) or in tactical situations (e.g., a fire team as they "shoot, move, and communicate"). And like all patterns in adversarial environments, they are contextual. Not every principle here is applicable to every adversarial domain. But the foundational principles of each domain are all represented. And if they hold in a general sense—across radically different configurations—then they have earned a place in this book.

## Sources

- Argyris, Chris. 1990. *Overcoming Organizational Defenses: Facilitating Organizational Learning.* Boston: Allyn & Bacon.
- Boyd, John. 1986. "A Discourse on Winning and Losing." Lecture notes, Maxwell AFB, AL.
- Clausewitz, Carl von. 1984. *On War.* Translated by Michael Howard and Peter Paret. Princeton, NJ: Princeton University Press.
- Freedman, Lawrence. 2013. *Strategy: A History.* New York: Oxford University Press.
- Hume, David. 1748. *An Enquiry Concerning Human Understanding.* London: A. Millar.
- Kahneman, Daniel. 2011. *Thinking, Fast and Slow.* New York: Farrar, Straus and Giroux.
- Kuta, Sarah. 2025. "'Robo-Bunnies' Are the Newest Weapon in the Fight Against Invasive Burmese Pythons in Florida." *Smithsonian Magazine,* July 21, 2025. http://www.smithsonianmag.com/smart-news/robo-bunnies-are-the-newest-weapon-in-the-fight-against-invasive-burmese-pythons-in-florida-180987018/
- Schelling, Thomas C. 1980. *The Strategy of Conflict.* Cambridge, MA: Harvard University Press.
- van Creveld, Martin. 1985. *Command in War.* Cambridge, MA: Harvard University Press.
- Von Neumann, John, and Oskar Morgenstern. 1944. *Theory of Games and Economic Behavior.* Princeton, NJ: Princeton University Press.

# PRINCIPLE 1
## PREVENT CONFLICT PROACTIVELY

*"An ounce of prevention is worth a pound of cure."* — Benjamin Franklin

I t might feel a little disappointing to open a book on conflict and immediately be told that the best strategy is to stop it from happening at all. Most of us want the drama of decisive engagements, the clever maneuver that turns the tide, the bold strike at just the right moment. Those chapters are coming. But we need to give prevention its due, despite how uninteresting it may seem at first glance. There is no rally, no sudden twist, no moment of glory. When prevention works, nothing happens. And that, as it turns out, is the point.

We need to preface this discussion—and, in fact, this entire book —by making an important point: the only reason conflict exists is because someone wants or needs something that impacts what someone else wants, with neither side conceding. Sometimes it's tangible—territory, resources, market share. Sometimes it's intangible—status, security, recognition, influence. Those wants and needs are real. You can't make them vanish by wishing them away. In fact, the challenge of proactive prevention lies in getting what you need and finding a way for the other side to get enough of what they need that they see no point in fighting. This is why prevention is an active process. It's not abstaining from ambition or avoiding competition—it's shaping the situation so both parties can satisfy core interests without turning to confrontation. If there were nothing you wanted, prevention would be easy. The work begins precisely because you do want something—and so does the other side—and sometimes those aims collide.

Before anyone chooses confrontation, they weigh the price of getting what they want by force. There are economics to conflict. Even a "win" carries bills: money and time burned, people harmed, legitimacy dented, and new obligations to police or occupy what was taken. Uncertainty fattens the tab—fog, friction, miscalculation, and escalation risk. Nature figured this out long ago. A big cat will usually keep the peace with dangerous prey not out of fear but because it knows the price of injury. It meets its dietary needs through easier prey or scavenging—still achieving its goal, just without risking a fight it doesn't need. Our own bodies operate the same way. Up to four out of five heart attacks and strokes are preventable. The solution isn't complicated—better food, more movement, early checkups —but prevention is almost invisible by design. The patient who never suffers a heart attack doesn't throw a parade for their arteries. They just keep living their life. The risk versus reward calculation for prevention almost always heavily favors it as the best strategy for achieving optimal return on investment. While it may not be the sexiest of options, supreme strategists recognize prevention's value and use it as a principle of first resort.

Prevention can take a few forms, including proactive prevention, avoidance, and deterrence. Proactive prevention is the superior of the three. *Avoidance* means you've already spotted trouble and choose to step away from it. That might keep you safe in the moment, but the underlying conditions remain. The seeds are still planted; you've just chosen not to be there when they sprout. *Deterrence* accepts that hostility exists and uses the threat of punishment to keep it contained: If you hit me, I'll hit you harder. Deterrence has its place, but it is reactive and brittle. It assumes the other side already wants to hurt you; it only argues about whether they dare. And if they ever believe you won't—or can't—make good on your threat, deterrence collapses. Proactive prevention moves the fight upstream, where the water is still calm. It's not about bracing for an attack—it's about shaping the environment so that an attack never makes sense in the first place. It removes the incentives to strike, builds relationships that make peace more profitable than war, and changes the terrain so that aggression feels irrational, expensive, or even self-destructive.

Across every type of adversarial struggle—whether it's between states, companies, teams, or individuals—proactive prevention comes down to six universal moves. These moves don't necessarily unfold in a strict sequence of time. Instead, they represent complementary levers—removing motives, blocking capabilities, building coopera-tion, reinforcing guardrails, shaping perceptions, and stabilizing the environment. Taken together, they form a complete prevention toolkit:

1. **Eliminate root incentives** – Remove the underlying reason the other side would see conflict as worthwhile. This is about neutralizing the motive to fight entirely, not just making it harder. If the "why" disappears, so does the strategic case for aggression. In business, that could mean redefining contested market boundaries so both companies can grow without poaching the same customers. In diplomacy, it could mean guaranteeing shared access to a disputed waterway so neither side gains from seizing it. Once the core incentive is gone, the fight no longer makes sense.

2. **Remove enabling conditions** – Accept that the other side may still want to fight, but make it physically, logistically, or technically impossible—or prohibitively costly—to execute an attack. This is about neutralizing capability or opportunity without relying on deterrent threats. In international relations, that might mean joint security patrols in contested waters that remove exploitable gaps.

In industry, it could mean controlling exclusive rights to a critical raw material or infrastructure chokepoint, leaving an adversary unable to act without your cooperation. The motive may remain, but the means to act are gone.

3. **Build mutual gains and interdependence** – Structure relationships so that both sides profit more from cooperation than they could from winning a conflict, even if the motive and means to fight still exist. This is about tilting the payoff equation so that breaking the relationship is self-destructive. Businesses might cross-license technology so each depends on the other's success. Sports leagues use revenue-sharing so top teams benefit when weaker teams survive and thrive. The choice not to fight comes from a rational calculation that cooperation outperforms confrontation.

4. **Reinforce protective structures** – Put in place agreed-upon mechanisms that can intercept disputes early, before they escalate into open conflict. This includes accessible outlets—formal or informal—where grievances can be aired and resolved at a lower level. In industry, that might be a standing arbitration panel for competitor disputes. In diplomacy, it could be a permanent joint committee empowered to resolve emerging issues before they spiral. In organizations, it could be an ombudsman or internal mediation process that addresses concerns before mistrust hardens. These structures give friction somewhere to go besides the battlefield.

5. **Shape perceptions early** – Establish your intentions, values, and boundaries before others define them for you. This is about controlling the interpretive frame in advance so that potential adversaries see peace as the rational choice. In cybersecurity, a well-run bug bounty program reframes would-be attackers as contributors, while public messaging avoids making you an attractive target. In

community relations, steady transparency and engagement build credibility long before it's tested. Once perceptions are shaped, they're harder to weaponize against you.

6. **Maintain environmental stability** – Keep the surrounding context stable enough that no one is forced toward desperate action. This is about managing systemic pressures so that external shocks don't create new incentives for conflict. For a multinational corporation, that could mean investing in the resilience of local communities to prevent unrest from disrupting operations. In a sports team, it might mean maintaining healthy player–management relations so internal rifts can't be exploited. Stability lowers the risk that dormant tensions ignite.

When seen together, these six moves form a complete prevention strategy. Move 1 kills the why by removing the motive to fight. Move 2 kills the can by stripping away the ability or opportunity to attack. Move 3 makes "won't" the best choice by making cooperation more rewarding than conflict. Move 4 catches sparks early by addressing grievances before they ignite. Move 5 controls the narrative by shaping perceptions in advance. Move 6 calms the whole landscape by keeping the wider environment stable enough that no one is driven toward desperate action.

Prevention doesn't look the same in every domain, but the underlying logic is constant: change the game so that the win conditions for your adversary no longer require your loss. Done well, proactive prevention is quiet, almost invisible. You won't get applause for the war that never happened, the lawsuit never filed, or the breach that never occurred. But those are victories all the same—and often the most valuable kind. From here, we'll look at two real-world cases: one where these principles transformed a high-risk relationship into a partnership, and another where the failure to apply them made war inevitable.

## Positive Example — Chief Dan Stump and the Springettsbury Township PD Turnaround

When Chief Dan Stump took command of the Springettsbury Township Police Department in Pennsylvania, the relationship between his officers and the community was fractured. Years of high-profile incidents—including a fatal officer-involved shooting in December 2012 —had deepened mistrust. That shooting, in which officers confronted Todd William Shultz, a shoplifter armed with knives and scissors, ended with Shultz's death, a federal lawsuit, and intense public scrutiny. The national conversation on police use of force was already tense; in Springettsbury, it had become personal.

Both sides had needs. The community wanted safety, respect, and assurance that their police acted with fairness and accountability. The department wanted legitimacy, the ability to do its work without constant suspicion, and an end to a cycle of confrontation and backlash. Both wanted peace—but not on terms the other fully trusted.

Stump recognized that these wants couldn't be met through avoidance. The tension was already there; simply stepping back from the public would not make it go away. Nor could deterrence solve it— warning the public not to interfere or criticize would only inflame the perception of an unaccountable force. What was needed was to break the cycle and insert *proactive prevention*: changing the environment so that future flashpoints would never push the relationship to the brink.

In February 2016, Stump courageously broke the cycle. He initiated a partnership with the federal Office of Justice Programs (OJP) Diagnostic Center. The aim was not just to patch over immediate problems, but to build systems that could address grievances before they hardened into hostility. This *eliminated root incentives* for conflict by removing the causes of mistrust and *reinforced protective structures* through external guidance and process reform.

He then invited community leaders to a public forum—an uncomfortable experience by his own admission—where he listened without defensiveness as residents voiced years of frustration. This openness began *shaping perceptions*, showing the department was willing to hear criticism and act on it. Stump went further, creating an advisory council made up of respected community members and granting them unprecedented access to the department's inner workings. This was *building mutual gains and interdependence*: the community gained influence over policing policy, and the department gained advocates with direct insight into its operations.

Over time, the changes took root. The department incorporated principles of procedural justice into daily practice, improved communication with residents, and implemented a stronger body camera policy—measures that removed *enabling conditions* for suspicion and escalation. Trust built slowly but steadily.

The proof came in a moment that would once have been combustible: the next officer-involved incident. Instead of outrage, community leaders spoke publicly in *support* of the department's handling of the matter. Potential adversaries had become stakeholders. In 2017, the department's work was recognized with the "Heart of Change" award from a local community association for bridging the police-community divide.

By addressing the underlying needs and wants of both sides— and creating ways to meet them without confrontation—Chief Stump demonstrated exactly what proactive prevention looks like in practice. The conditions for future conflict were dismantled before they could take shape, and in their place, a framework for cooperation was built.

### Negative Example — Britain's Falklands Miscalculation

For decades, the Falkland Islands (Las Malvinas to Argentines) simmered as a low-level diplomatic dispute between Britain and Argentina. It was a known tension but not an urgent one—the kind of unresolved issue that both sides learned to live with. Beneath the surface, though, the two nations' needs and wants were pulling in opposite directions.

For Argentina's ruling military junta in the early 1980s, domestic conditions were collapsing. The economy was in free fall, inflation was out of control, and public unrest was growing. The generals needed a unifying cause—something to rally national pride and distract from internal failures. Recovering the Falklands offered exactly that: a patriotic prize that could be claimed quickly and, they believed, at relatively low cost.

Britain's wants were simpler but no less important. The government wanted to maintain sovereignty over the islands, protect the rights and security of its citizens there, and avoid the financial and political drain of a far-off conflict. What it did not want was to spend heavily on a distant outpost with limited strategic value in peacetime.

This was a moment when proactive prevention could have worked. The UK could have recognized the growing incentive for the junta to act and sought ways to meet Argentina's need for prestige and legitimacy without surrendering sovereignty. Eliminating root incentives could have meant offering expanded fishing rights, joint resource management, or a framework for shared economic projects involving the islands—ways for the junta to claim a visible "win" without war.

Instead, London's actions sent the opposite signals. In 1981, as part of a defense review, Britain decided to withdraw its only permanent naval presence in the South Atlantic, the ice patrol ship *HMS Endurance*. To budget officials, this was a minor cost-saving measure. To Buenos Aires, it looked like removing a protective structure—a sign that Britain's willingness to defend the islands was in doubt.

Around the same time, the British Nationality Act of 1981 restructured nationality categories in ways that weakened some Islanders' formal ties to Britain. The change signaled that London's commitment might be waning.

British diplomats—including Foreign Secretary Lord Carrington —warned that such moves would be interpreted in Argentina as reduced commitment. Parliamentary debates after the war would assess that such moves had been, "almost a green light to the Argentine Government to invade." Yet, at the time, budgetary and other priorities carried the day, and the warnings of Carrington and others went unheeded.

Diplomatic initiatives, such as "leaseback" proposals and UN-mediated talks, were intermittently pursued but never matured into a stabilizing arrangement. By early 1982, Argentina's junta assessed the islands as a low-risk, high-reward opportunity. On 2 April, Argentine forces invaded. From their perspective, the decision was rational: Britain had signaled disinterest, removed visible defenses, and failed to create alternative pathways to meet Argentina's domestic needs.

The war that followed lasted ten weeks, cost over 900 lives, and inflicted heavy political and economic damage. Britain ultimately regained control of the islands, but at a cost far greater than any proactive measures would have required.

The Falklands conflict is a reminder that when the other side's needs and wants are ignored, you leave them to find their own path to fulfillment. If that path leads through you—and you've done nothing to make peace more rewarding than war—you may have already lost before the first shot is fired.

## Principles in Action — Reflection Questions

1. Have we clearly identified what we want, what the other side wants, and the space where both can get enough to make confrontation unnecessary?
2. What deliberate, ongoing steps are we taking to remove incentives for conflict, close off enabling conditions, and build interdependence that rewards cooperation?
3. Which preventive investments—in relationships, governance processes, technical systems, or agreements— would most reduce the likelihood of conflict, and are they properly resourced right now?
4. How are we actively shaping how others see our intentions and resolve, so that our posture reduces—not provokes— the appeal of conflict?

5. Are we monitoring the broader environment for changes that could push others toward desperate action, and moving early to stabilize those conditions?

## Sources

- Akins, Joyce. 2010. "The Falklands War: A Study in Crisis Management and Public Diplomacy." *Naval War College Review* 63, no. 3: 97–116.
- Associated Press. 2017. "Case in Fatal Shooting of Kmart Shoplifter Settled for $285K." *The Washington Times,* April 20, 2017.
- BBC News. 2007. "Timeline: The Falklands Conflict." *BBC News,* April 2, 2007.
- Britannica. n.d. "Falkland Islands War." *Encyclopedia Britannica.*
- CBS 21 News. 2019. "Familiar Face Now in Charge at Springettsbury Township Police." July 16, 2019.
- Centers for Disease Control and Prevention. 2024. *Fast Facts: Health and Economic Costs of Chronic Conditions.* Atlanta, GA: CDC.
- CNN. n.d. "Falklands War: Fast Facts." *CNN.*
- Franklin, Benjamin. 1735. "An Ounce of Prevention Is Worth a Pound of Cure." *Pennsylvania Gazette,* February 4, 1735.
- Franks, Lord. 1983. *Falkland Islands Review: Report of a Committee of Privy Counsellors.* London: HMSO, January 1983.
- Freedman, Lawrence. 2005. *The Official History of the Falklands Campaign, Vol. 1: The Origins of the Falklands War.* London: Routledge.
- Hansard. 1982. *House of Commons Debate, Falkland Islands,* April 7, 1982, vol. 21, cc961–1022. London: UK Parliament.

- "Falkland Islands – The British Nationality Act of 1981." n.d. *Hansard,* UK Parliament.
- "Falklands War: By the Numbers." n.d. *The Telegraph.*
- The Guardian. n.d. "The Falklands War: 30 Years On." *The Guardian.*
- International Association of Chiefs of Police. 2017. "How Small Agencies Use Data-Driven Decision-Making." Session #1962, IACP Annual Conference.
- Jacob Litigation. 2014. "Estate of Todd W. Shultz et al. v. Gregory T. Hadfield et al." December 17, 2014.
- Office of Justice Programs Diagnostic Center. 2019. *Engagement Summary: Springettsbury Township Police Department 2016–2018.* Washington, DC: U.S. Department of Justice.
- PennPRIME. 2018. "Department Excellence Honored at PennPRIME Conference." http://www.pennprime.com/ news/department-excellence-honored-at-pennprime-conference/
- Springettsbury Township Board of Supervisors. 2019. Meeting minutes, March 28, 2019.
- Wainwright, C. Martin. 2017. *Crispus Attucks York 2016 Annual Report.* York, PA: CAY.
- World Heart Federation. 2024. *CVD Prevention.* Geneva: WHF.

# PRINCIPLE 2
## DEFINE VICTORY WITH PRECISION

*"No one starts a war—or rather, no one in his senses ought to do so—without first being clear in his mind what he intends to achieve by that war and how he intends to conduct it."* — Carl von Clausewitz

Who can forget the image of President George W. Bush standing beneath a giant *Mission Accomplished* banner on the flight deck of the USS *Abraham Lincoln* in May 2003? From a public-relations standpoint, the scene was masterful: the flight-suit landing, the roaring jet, the confident declaration that "major combat operations in Iraq have ended." But from a strategic standpoint, it was catastrophic. The United States had not, in fact, defined an achievable end state for its involvement in Iraq. There was no shared sentence, agreed upon by the White House, the Pentagon, the State Department, and coalition partners, describing exactly what would constitute "victory" and when it would be deemed complete. The result was many more years of shifting objectives, spiraling costs, and mounting casualties. The episode endures as a cautionary emblem of what happens when leaders celebrate the finish line before they have even measured where it lies.

Principle 2 exists to prevent that mistake. It demands that an actor contemplating conflict—whether a nation planning a campaign, a corporation preparing litigation, or a cybersecurity team considering counterstrikes—begin by writing down a single, unambiguous statement of success. That statement must be articulated in a way that anyone, friend or foe, could read it after a conflict and declare definitively whether it was a success or failure. Anything less invites drift, and drift is fatal because conflict amplifies every tendency toward distraction, emotion, and overreach.

The most reliable test for whether a proposed outcome is sufficiently clear is the SMART standard: Specific, Measurable, Achievable, Relevant, and Time-Bound. Although first popularized in management circles, SMART applies across every arena where adversaries clash—from military campaigns to medicine, from political contests to invasive species eradication.

- **Specific** means naming the exact end state you intend to reach, in clear, concrete terms.
  - *Meets*: In armed conflict, *end hostilities by securing a signed peace agreement that restores prewar borders.*
  - *Fails*: In armed conflict, *weaken the enemy's position.*
- **Measurable** means success can be confirmed by an outside observer without insider knowledge.
  - *Meets*: In political campaigns, *win at least 270 electoral votes in the U.S. presidential election.*
  - *Fails*: In political campaigns, *run a strong campaign.*
- **Achievable** means the goal is realistic given time, resources, and constraints. Ambition is fine; impossibility is not.
  - *Meets*: In martial arts competition, *place top three in the regional Brazilian Jiu-Jitsu championship in the featherweight division within two years.*
  - *Fails*: In martial arts competition, *become the undisputed best fighter in the world in all weight classes within a year.*
- **Relevant** means the outcome directly serves the larger

strategic purpose, rather than drifting into unrelated victories.

- ○ *Meets:* In invasive plant control, *stop the spread of kudzu within the state to protect native hardwood forests*, if the purpose is preserving biodiversity.
- ○ *Fails:* In invasive plant control, *eliminate all non-native species everywhere*, when only kudzu poses the urgent threat.
- **Time-Bound** means setting a clear deadline that drives action and prevents drift.
  - ○ *Meets:* In medicine, *develop and deploy a safe, effective vaccine for Pathogen X within 18 months*.
  - ○ *Fails:* In medicine, *develop a vaccine for Pathogen X*.

SMART in this context is not a performance-tracking tool for small milestones. It exists to lock in the one sentence that defines exactly where the campaign must end—the condition that, when met, signals the mission is over. Everything else—intelligence cycles, resource allocations, morale efforts—exists to march toward that sentence.

When leaders neglect this discipline, several predictable pathologies emerge. Scope expands as new desires piggy-back onto the mission. Stakeholders begin rewriting victory to match their own incentives. Emotions hijack decision-making, encouraging retaliation for its own sake. Budgets hemorrhage because no one can say which expenditure is still essential. Adversaries exploit ambiguity to lure their opponents into side contests. Sub-units, afraid to breach invisible limits, become timid. And, finally, withdrawal becomes politically impossible because no one can pinpoint the moment when the campaign is "over," so promises of closure ring hollow. Morale is thereby critically affected.

For purpose of clarity, there must be clean separation between vision, mission, and victory condition. Vision articulates the moral horizon ("a world safe from coercion"). Mission explains the enduring role ("maintain collective defense"). Victory condition belongs to a particular campaign ("return a crewed spacecraft safely from the lunar surface by December 31, 1969"). When these layers blur, operations lose traction.

Precision of victory condition is universal across every axis of UTAD. The polarity of a confrontation—whether it is AI vs. AI or human vs. virus—does not alter the necessity of a crystal-clear end state. In each case, systems under stress obey the gradient of definition: vague goals invite entropy; precise goals funnel energy toward closure.

Finally, a SMART outcome serves as a predeclared threshold separating prevention from engagement. So long as the outcome can be attained through non-violent means—diplomacy, incentives, or environmental shaping—restraint is prudent. When an adversary action makes the outcome unattainable without force, escalation becomes legitimate because the threshold has been crossed by objective criteria, not by emotional impulse.

**Positive Example — Apollo 11: A Goal as Clear as the Night Sky**

The Cold War was as much psychological theatre as geopolitical rivalry. After the Soviet astronaut Yuri Gagarin orbited Earth in April 1961, the United States needed a visible, decisive reply. President John F. Kennedy delivered one in a single breath before Congress on May 25, 1961: "I believe that this nation should commit itself to achieving the goal, before this decade is out, of landing a man on the Moon and returning him safely to the Earth." The sentence met every element of SMART.

It was **Specific:** one crewed landing, one safe return.

It was **Measurable:** footprints on lunar dust, splash-down in the Pacific—binary.

It was **Achievable:** a stretch, yet grounded in Mercury and Gemini mission data.

It was **Relevant:** winning the perception war against the Soviet Union.

It was **Time-Bound:** December 31, 1969.

That clarity reorganized the world's most complex engineering effort into a convergent campaign. NASA leaders used the statement as a litmus test for every subsystem. If a component did not shorten the path to that lunar landing or reduce the odds of astronaut loss, it was redesigned or scrapped. When the Apollo 1 fire killed three astronauts in 1967, the program nearly collapsed. But the destination never shifted, and because it did not, the recovery did not require renegotiating purpose—only revalidating methods. Had the objective been "demonstrate space superiority," the tragedy might have shattered unity, but the lunar imperative held.

On July 20, 1969, Neil Armstrong stepped onto Mare Tranquillitatis; four days later, the crew splashed down back on Earth safely. Kennedy's deadline still had five months to spare. The achievement remains unmatched in its clarity of purpose: a campaign that began with a clear victory condition ended with an indelible photograph of a boot print.

### Negative Example — The War on Drugs: A Campaign Without a Victory Condition

The late 20th century was a period of intense social anxiety in the United States over crime and illicit drug use. The government needed a visible, decisive reply. President Richard Nixon delivered one in a fateful declaration on June 17, 1971: "America's public enemy number one in the United States is drug abuse. In order to fight and defeat this enemy, it is necessary to wage a new, all-out offensive." This launched the "War on Drugs," but unlike a clear strategic objective, this mission was defined by a vague moral vision—a "drug-free society"—rather than a precise, achievable outcome. The sentence failed every element of SMART.

It was **Not Specific:** "A drug-free society" is a nebulous concept, not a concrete state of the world. It fails to name a quantifiable or achievable condition.

It was **Not Measurable:** There was no verifiable metric for success. Instead, the campaign measured activity—arrests made, drugs seized—which are inputs, not outcomes. A neutral observer could not declare victory.

It was **Not Achievable:** The total eradication of drug use from a complex society is a sociological fantasy, ignoring the deep roots of addiction and demand. Strategy must respect the constraints of human nature.

It was **Not Relevant:** It targeted a symptom rather than identifying root causes of drug use. Over time, the tactics deployed, especially mass incarceration, created secondary social crises that were arguably more destructive than the original problem, disconnecting the effort from its supposed purpose of creating a healthier society.

It was **Not Time-Bound:** The "war" was declared without an end date, creating a permanent conflict rather than a campaign with a deadline.

∼

THIS WAS NOT JUST A VAGUE VISION—IT WAS A CAMPAIGN WITHOUT AN outcome sentence, and the pathologies predicted earlier unfolded almost exactly as described. The ultimate indictment of this undefined war is not just its failure to achieve its goal, but its correlation with a dramatic *worsening* of the problem. A look at the data reveals that after more than five decades and over a trillion dollars spent, the nation is not closer to being "drug-free"—it is drowning in a more complex and deadly drug crisis than the one that existed in 1971.

Consider the most definitive metric: human lives lost.

- In 1971, the United States recorded about 6,800 drug overdose deaths (~3.5 per 100,000).
- In 2023, CDC reported 105,007 overdose deaths (~31.3 per 100,000).

Adjusted for population growth, the death rate from drug overdoses is now nearly ten times higher than it was at the start of the "war."

The scale of drug use has also exploded. In the early 1970s, lifetime use of any illicit drug other than marijuana was substantially lower than today. By contrast, the 2022 National Survey on Drug Use and Health found that over 70 million Americans—more than one in five people aged 12 or older—had used an illicit drug in the past year.

The pathologies predicted by the chapter's principles are borne out by these statistics. The ambiguous mission did not just lead to hemorrhaging budgets and scope creep; it created a strategic vacuum. While the "war" focused on arrests and punishment (inputs), the adversary—the complex network of addiction, demand, and trafficking—adapted, evolved, and grew stronger. The rise of synthetic opioids like fentanyl is in part a consequence of this strategic failure, as traffickers adapted to enforcement by shifting toward more potent, compact drugs the original 'war' was never designed to fight.

Decades later, the campaign has no end in sight. The achievement is not a photograph of mission accomplished, but a tragic legacy of mass incarceration and a public health crisis that has spiraled out of control. The War on Drugs remains unmatched in its clarity of failure: a campaign that began with a sentence promising victory ended with a nation suffering more drug-related death and despair than ever before.

## Principles in Action — Reflection Questions

1. Have you written a single-sentence conflict outcome measure that a neutral observer could evaluate without insider context?
2. Do all principal stakeholders endorse that exact sentence in writing, and is the document still current?
3. What indicator, tracked daily, will warn leaders that the outcome is drifting beyond "Achievable," and who is responsible for acting on it?
4. If ordered to disengage tomorrow, could you declare victory—or defeat—without rewriting the outcome?

5. Have you documented the precise condition under which the current outcome becomes unattainable through prevention alone, so subordinates need not invent their own red lines?

## Sources

- Clausewitz, Carl von. 1984. *On War*. Translated by Michael Howard and Peter Paret. Princeton, NJ: Princeton University Press.
- Council on Foreign Relations. n.d. "Timeline: The Iraq War." *Council on Foreign Relations.*
- Doran, George T. 1981. "There's a S.M.A.R.T. Way to Write Management's Goals and Objectives." *Management Review* 70, no. 11: 35–36.
- Drug Policy Alliance. 2015. "The War on Drugs, A Trillion Dollar Failure." News release, January 8, 2015. http://drugpolicy.org/news/2015/01/war-drugs-trillion-dollar-failure
- Kennedy, John F. 1961. "Special Message to the Congress on Urgent National Needs." Speech, May 25.
- Musto, David F. 1999. *The American Disease: Origins of Narcotic Control.* 3rd ed. New York: Oxford University Press.
- National Academies of Sciences, Engineering, and Medicine. 2017. *Pain Management and the Opioid Epidemic: Balancing Societal and Individual Benefits and Risks of Prescription Opioid Use.* Washington, DC: The National Academies Press. https://doi.org/10.17226/24781
- NASA. n.d. "Apollo 11 Mission Overview." National Aeronautics and Space Administration. http://www.nasa.gov/mission_pages/apollo/missions/apollo11.html
- Nixon, Richard. 1971. "Remarks About an Intensified Program for Drug Abuse Prevention and Control." June 17, 1971. *The American Presidency Project.* http://www.

presidency.ucsb.edu/documents/remarks-about-intensified-program-for-drug-abuse-prevention-and-control

- White House Office of the Press Secretary. 2003. "President Bush Announces Major Combat Operations in Iraq Have Ended." May 1, 2003.

# PRINCIPLE 3
## BUILD CAPABILITY THAT RENDERS STRATEGY IRRELEVANT

*"God is on the side with the best artillery."* — Military Aphorism

When most people think about winning a conflict, they picture strategy taking center stage—the cunning battle plan, the market move that catches a rival sleeping, the brilliant counter in the final round. History celebrates those moments. But they matter most when the two sides already have *roughly* equal strength. Where there's a large gap in capability—troops and firepower, manufacturing scale, computing power, biological resilience—the stronger side usually wins, and strategy mainly speeds up or slows down the inevitable. As the old football saying goes, "it's more about the Jimmies and the Joes than the Xs and the Os."

Military historian John Keegan recounted centuries of scouts, cryptographers, and spymasters in *Intelligence in War*. His conclusion was stark: intelligence—by which we can also mean for our purposes here any lever other than raw capacity—shapes outcomes only when fighting power is balanced. Fighting power, he wrote, is the mistress of the warrior; intelligence is but the handmaiden. To put it more plainly: Capability is of primary importance. Strategy is of secondary importance.

The primacy of capability holds true across domains. In human-to-human contests, it's muscle, morale, and materiel; in AI-to-AI, parameter count, training data, and inference hardware; in unconscious life struggles, mutation rate, biochemical arsenal, replication speed. A cunning bacteriophage may try alternate infection paths, but if its replication speed can't outpace host immunity, the campaign ends quickly. The principle is the same: strategy operates inside a ceiling set by capability. Build enough of it—physical, digital, financial, biological—and the contest stops being chess and becomes demolition.

## The Three Strands of Capability

1. **Innate Capacity.** The raw material—genetics, geography, resources, or other built-in advantages that exist before training begins. The NBA will sign a seven-foot prospect who has never played organized basketball because height can't be taught; skills can. In predator–prey dynamics, a cheetah's top speed comes from anatomy, not practice. Selecting, recruiting, and positioning for superior raw capacity grants advantages rivals cannot match through technique alone.

2. **Preparation.** Preparation optimizes innate capacity. It's the countless sparring rounds, live-fire rehearsals, red-team exercises, and manufacturing quality loops that make strength reliable under pressure. The fastest sprinter loses to a slightly slower but better-conditioned rival if they gas out halfway. Preparation compounds over time, building reflexes, endurance, timing, and calm—the parts of capability the audience never sees but always feels when it matters.

3. **Tools.** The right tools and systems multiply effort beyond what raw capacity and training alone can deliver. In war, undisciplined soldiers with repeating rifles overcome even the fiercest and most skillful archers. In medicine, antibiotics changed infection outcomes regardless of a provider's poor bedside manner. In Formula 1, the most skilled driver cannot win in a slow car. In cyber, purpose-built hardware and hardened pipelines unlock classes of defense that no amount of "try harder" can substitute for. Tools don't replace skill or innate gifts, but they decisively tilt the balance when margins allow it.

## Sustaining Capability

When all three strands are strong and regularly renewed, capability becomes the singular force that outplays strategy. When one frays, the rope can snap the moment pressure peaks. To avoid fraying, sustainment is key. Sustainment must keep pace with the competitive environment. In low-change arenas, a five-year modernization cycle may suffice; by contrast, in cyberwarfare, the upgrade window is weeks or even days. Capability demands reinvestment at the environment's pace.

Internal alignment also matters for sustainment. Organizations must value drills, maintenance windows, education budgets, and quiet R&D—cost sinks until the day they pay out. The U.S. Navy's nuclear program, built on relentless procedural compliance, has suffered no reactor casualties in six decades. BP's Macondo well, by contrast, skipped cement bond logs, cut mud circulation, and ignored equipment shortages—fraying preparation until Deepwater Horizon blew, costing over sixty billion dollars. Culture decides whether capability sustains or decays.

## Systemic Advantage of Capability

**Capability rearranges time.** Strength buys deliberation while compressing the clock for your opponent. The USS Missouri off Okinawa could let kamikaze flights appear on radar and still respond with layered defenses. Modern incident-response teams isolate malware in milliseconds, forcing attackers to burn new exploits. The strong choose when to decide; the weak are hurried into mistakes.

**Capability shrinks terrain.** Genghis Khan perfected horse-bow unit synergy so riders covered sixty miles a day; distance that once sheltered kingdoms became a morning commute. Online, a ten-thousand-node botnet turns the globe into a single strike platform. Territory—physical or digital—belongs to those who can traverse and exploit it at speed.

**Capability confers emotional advantage.** Confidence is a tactical resource. At San Juan Hill, the Buffalo Soldiers kept moving under fire because they trusted their gear—reliable repeating rifles in their hands and machine-gun support behind them. Answering enemy Mauser fire in that situation felt possible. The same logic shows up in court: a lawyer who has every document and fact at hand stays calm while the other side guesses. That steadiness isn't bravado; it comes from capability—the grounded belief that your side can handle the worst moments.

**Capability decides exit options.** The capacity to leave makes engagement voluntary. A firm with a year of cash runway negotiates with a potential buyer as a peer, not a supplicant. An army with secure supply lines can disengage to fight another day; a besieged force must gamble or starve. A job seeker with scarce savings might accept a toxic workplace; one with six months living expenses can wait for better fit.

**Capability supports ethical decision-making.** Superior strength can widen the range of humane or ethical choices. A public-health system with ICU surge capacity can adopt less harsh quarantine measures. A solvent firm can disclose breaches quickly; a fragile one may hide them. Power can afford patience and transparency; weakness often breeds corner-cutting and escalation.

$\sim$

Taken together, these perspectives harden into a simple conviction (and please don't misinterpret the phrase): *invest in Keegan's "mistress."* Strategy will always matter—especially when near-equals meet—but it pales in comparison to brute strength that is refined by preparation and equipped with superior tools. In the examples that follow, you'll see the contrast up close—first in a capability leap so large it reshaped an entire discipline, and then in a slow decay that toppled a champion at the bell.

### Positive Example — AlphaGo's Leap Beyond Human Capability

For centuries, the game of Go was the canonical test of human strategic capability. A 19×19 board admits on the order of $2 \times 10^{170}$ *legal* positions. To put that number in perspective, if we took every atom in the observable universe and gave it its own corresponding universe, and then counted *all* the atoms in *all* the resulting universes, it would *still* be orders of magnitude *less* than the possible configurations in Go. This reality renders brute-force strategies impossible. Masters therefore relied on instinct refined over decades, passing down *joseki* and *fuseki* patterns like craft recipes. Then, in 2016, DeepMind's AlphaGo arrived. The game would soon be forever changed.

From the beginning, the DeepMind team built immense capability across all three strands:

**Innate Capacity.** First, they secured vast computing power—the digital muscle that let the system explore and evaluate far more possible move sequences than any human could.

**Preparation.** They first learned from millions of expert positions, then switched to relentless self-play, generating more high-quality experience in weeks than humanity had accumulated in millennia.

**Tools.** Not just more power, but better *use* of power: chips built for the math, a training/search pipeline that proposed good moves and evaluated positions quickly, and a search method that spent cycles only where it counted—turning raw horsepower into fast, reliable strength.

On the competition stage, the result was unmistakable. In March 2016 AlphaGo defeated 18-time world champion Lee Sedol 4–1. This was just the beginning. In early 2017 the "Master" version ran up a 60–0 streak online against top Go professionals, and then beat world champion Ke Jie by a score of 3–0 at the *Future of Go Summit*; AlphaGo retired from competition the same week, being replaced by the more-powerful AlphaGo Zero.

At this stage, human strategies had *already* been rendered irrelevant. In fact, many of the principles you will encounter in this book—gathering intelligence, exploiting positioning, timing, relentless attacking, deception—even if employed perfectly, would do little but delay an inevitable AI victory. That's what we mean when we say, "build capability that renders strategy irrelevant." But DeepMind's AI wasn't done yet. Not even close.

The internal capability curve kept bending upward. In October 2017, AlphaGo Zero—trained *from scratch*, no human games—defeated the earlier, champion-beating AlphaGo 100–0. Weeks later, AlphaZero generalized the same approach to Go, chess, and shogi, reaching superhuman performance from random play in about a day. In 2020, the MuZero evolution pushed further, planning effectively *without even being told the rules*, and matching or exceeding AlphaZero's Go results. These evolutions represent new capability plateaus, each one rendering yesterday's hard-won schemas obsolete.

The shockwave hit human play immediately. Pros began studying AI self-play libraries; shapes once scorned became standard, like frequent early 3-3 invasions against 4-4 corners. Today, top training partners are engines rather than humans, and publicly available systems such as KataGo give players access to strength that was science fiction a decade ago.

One nuance matters for readers who follow Go closely: in 2023 researchers showed that carefully crafted adversarial tactics can lure state-of-the-art Go AIs (including KataGo) into bizarre blunders—and even enabled an amateur to beat a top bot repeatedly. Those are targeted exploits, not "normal play," but they're a healthy reminder that capability can *sometimes* be overcome when a critical vulnerability is discovered. This is, however, the exception, and the broader story stands: since 2017 no human has beaten the strongest Go AIs in normal play, and the successors have only grown exponentially stronger.

AlphaGo wasn't a clever trick that outthought humanity; it was a deliberate build across *innate capacity*, *preparation*, and *tools* that lifted the ceiling of what was possible. Once that ceiling rose, strategy itself had to reorient around the new terrain.

### Negative Example — Mike Tyson's Collapse Against Buster Douglas

In February 1990, Mike Tyson entered the Tokyo Dome as the undefeated, undisputed heavyweight champion (37–0, 33 KOs). He wasn't merely winning—he was ending world-class fights in minutes. Buster Douglas, a capable but inconsistent contender, was a 42-to-1 underdog. Almost everyone expected another quick Tyson knockout.

From the opening bell, Douglas set the pace with a long, disciplined jab and steady footwork. Tyson, who once slipped jabs as if on a spring, wasn't moving his head the same way. Swelling gathered around his left eye and, between rounds, his corner struggled to control it. By the middle rounds Douglas was landing clean, long-range combinations; Tyson had moments in close but couldn't sustain pressure. In Round 8 Tyson finally landed a heavy shot that dropped Douglas—proof the old power was still there—but Douglas beat the count, steadied himself behind the jab, and took the next rounds. In the 10th, a clean four-punch sequence finished Tyson—mouthpiece askew, aura gone.

This upset is best understood as a *capability failure* with one strand intact and two strands missing.

**What was intact — Innate Capacity.** Tyson's natural gifts remained: rare speed, explosive power, and toughness. The Round 8 knockdown is the clearest evidence—on talent alone, he could still change a fight with one punch.

**What was missing — Preparation.** The training discipline that made those gifts reliable had eroded. After parting with trainer Kevin Rooney, the old routine—hard roadwork, intense sparring, and daily drilling of head movement and compact combinations—slipped. In the ring that night, the tells were obvious: fewer slips and weaves against the jab, slower feet, visible fatigue by the middle rounds.

**What was missing — Tools.** The basic between-rounds system that keeps a fighter functional wasn't in place. When swelling built around Tyson's eye, his corner lacked the standard cold-metal tool used to press swelling down and improvised with a latex glove full of ice water. That's not a small detail: if you can't control swelling, your fighter can't see the jab—exactly the punch Douglas used to control distance all night.

Tyson's loss wasn't a mystery and it wasn't a miracle. His *innate capacity* was still there; his *preparation* and *tools* were not. That gap turned a fight he should have managed into a fight he couldn't control. We can summarize this example by simply inverting Principle 3: *his lack of capability rendered the opposition's strategy relevant.* Talent can get you to the fight, but you need all three strands to win it.

### Principles in Action — Reflection Questions

1. Are we actively expanding all three strands—access to innate capacity (talent, compute, resources), preparation, and tools—or coasting on legacy strength?
2. Which emerging competitor or technology could tilt time or terrain against us, and where are we over-investing now to keep that tilt from materializing?
3. Where do audits reveal silent decay—aging platforms, skill erosion, data drift, or under-equipped support

systems—and who has the mandate and budget to trigger aggressive replenishment?

4. After major successes, have we scheduled the next capability-building cycle (people, training, tooling) at a fixed cadence, or does celebration quietly mark the start of complacency?

5. Can we name a rival whose capacity upgrades tomorrow would render our best strategies marginal—and what is our preempt plan (build, buy, ally, or exit)?

## Sources

- DeepMind. 2016. *AlphaGo: Mastering the Game of Go with Deep Neural Networks and Tree Search.* White paper. London: DeepMind Technologies.
- Douglas, Buster, and Randy Gordon. 2018. *42 to 1: The Story of the Biggest Upset in Boxing History.* B&G Publishing.
- Keegan, John. 2004. *Intelligence in War: Knowledge of the Enemy from Napoleon to Al-Qaeda.* New York: Vintage Books.
- Schrittwieser, Julian, et al. 2020. "Mastering Atari, Go, Chess and Shogi by Planning with a Learned Model." *Nature* 588: 604–609. https://doi.org/10.1038/s41586-020-03051-4
- Silver, David, Aja Huang, Chris J. Maddison, et al. 2016. "Mastering the Game of Go with Deep Neural Networks and Tree Search." *Nature* 529: 484–489. https://doi.org/10.1038/nature16961
- Silver, David, Julian Schrittwieser, et al. 2017. "Mastering the Game of Go without Human Knowledge." *Nature* 550: 354–359. https://doi.org/10.1038/nature24270
- Silver, David, et al. 2018. "A General Reinforcement Learning Algorithm That Masters Chess, Shogi, and Go

through Self-Play." *Science* 362: 1140–1144. https://doi.org/10. 1126/science.aar6404

- Sports Illustrated. 1990. "Tyson Shocker: The Night the Myth Exploded." *Sports Illustrated,* February 19, 1990.
- Tromp, John. 2016. "Number of Legal Positions in Go." John Tromp's Homepage. Last modified January 2016. http://tromp.github.io/go/legal.html
- Tyson, Mike, and Larry Sloman. 2013. *Undisputed Truth.* New York: Blue Rider Press.
- Wang, Tony Tong, et al. 2023. "Adversarial Policies Beat Superhuman Go AIs." In *Proceedings of the 40th International Conference on Machine Learning (ICML 2023).* arXiv:2211.00241. https://arxiv.org/abs/2211.00241

# PRINCIPLE 4
## KNOW THE ADVERSARY BETTER THAN THEY KNOW YOU

*"It is said that if you know your enemies and know yourself, you will not be imperiled in a hundred battles."* — Sun Tzu

This principle sits closest to my heart. After nearly three decades in the intelligence profession I have watched fortunes rise and fall on a single variable: who sees the board more clearly and hides their own pieces more skillfully. Capability—the "mistress," as John Keegan reminds us—decides contests when strength is significantly unequal. But once something approaching parity returns, insight becomes the steering wheel of power. In that parity zone, the side that understands the adversary's habits, motives, thresholds, and blind spots gains leverage out of proportion to all else. Think of capability as the hammer—blunt power that can break stone through sheer force. Intelligence is the hand that guides it. A mis-swung hammer can still do damage, but the sculptor with the sharper eye shapes outcomes with far greater precision. The value of intelligence is even seen in nature: meerkats post sentinels that spend up to 20% of daylight scanning for hawks; one chirp can mean the difference between life and death.

Knowing the adversary is not passive monitoring; it is structured, purposeful, lifecycle intelligence. Collection begins well before open conflict, shadows the opponent through every phase of engagement, and continues afterward to harden lessons. Done well, it serves two intertwined aims:

1. **Targeted Collection with Valid Analysis.** Identify, acquire, and correctly interpret information relevant to the adversary's capabilities, intentions, posture, and likely courses of action. This includes both understanding their current state and forecasting probable developments so you can prepare countercapabilities, identify vulnerabilities, and anticipate shifts. In formal tradecraft, this aim spans the intelligence cycle: planning and direction, collection, exploitation, analysis, and dissemination. Executing it well means gathering the right data at the right time, enforcing analytic discipline to avoid mirror-imaging and confirmation bias, fusing multiple sources, and protecting dissenting voices so harmful assumptions don't calcify.

2. **Denial of Adversarial Intelligence Collection.** Actively prevent the adversary from gathering accurate intelligence about you at tactical, operational, and strategic levels. The core objective is to deny them a reliable picture: control emissions, sanitize communications, compartmentalize, employ decoys, and add operational noise that degrades their signal. Deception can contribute, but it is covered in depth in another chapter; here the emphasis is on obstructing their collection and corrupting their sensing, not on elaborate ruses.

～

THESE AIMS REINFORCE EACH OTHER IN A FEEDBACK LOOP. EVERY correct insight enables shaping actions that generate better subsequent collection. Every successful denial deprives the adversary of calibration points, increasing the odds that their next move is misjudged. The better you see, the easier it is to see more and to remain unseen. The less they see, the more their own moves drift into error. Compound effects build over time.

Real intelligence work is slow and deliberate, not a cinematic epiphany where an overworked and hard-drinking analyst discovers a single piece of evidence that turns the tide of the war. Months of signals may be needed to confirm a change in fuel rations. A single uncorroborated human-source report may be kept in play if its reasoning is sound and it aligns with other known facts. A spreadsheet of chassis serial numbers may betray a hidden redeployment. Fusion between good collection and good analysis is key. Analysts without tradecraft miss idiosyncratic cues; collectors who ignore context bring home noise. Only disciplined fusion turns fragments into foresight.

Denial should be likewise deliberate, because while you labor to know them, they labor to know you. Every conference talk, hiring blip, procurement tender, and code commit is a tile in their mosaic. Intelligence work is therefore two-sided: one track studies the foe; the other scrubs your own exhaust. Insider-threat monitoring, need-to-know access, red-teamed public disclosures, and silent product changes that mask true capability deny the opponent easy wins.

Another factor to consider in intelligence collection is *ethics*. Knowledge tests restraint. Deep collection exposes private vulnerabilities—family, faith, reputation. Exploiting those cracks may hasten victory but corrodes integrity. That tension previews a later discussion on moral coherence: power that eats conscience decays from within. Strong actors keep red-line rules even when no referee watches.

When both aims are executed well, modest forces can outmaneuver larger ones. When either aim fails, even superior capability can be ambushed by surprise. The case studies that follow show both outcomes in crisp relief.

## Positive Example — Holly Holm vs. Ronda Rousey

Let me acknowledge the elephant in the octagon: yes, this is our second combat-sports story in as many chapters—Tyson bowed out just moments ago. Blame the data set, not the author; when it comes to intelligence asymmetry, few modern events put the lesson in 4-oz gloves as cleanly as Holly Holm's dismantling of Ronda Rousey.

Who hasn't heard of Ronda Rousey, one of the better-known darling children of UFC? Rousey entered Melbourne undefeated, finishing her previous three opponents in an average of just *21 seconds*. Footage suggested a terminator: blitz forward, clinch, hip toss, armbar, collect bonus. But footage also whispered patterns. As I often tell my students in the intelligence program at Florida State University: *patterns breed vulnerabilities*. Holm's coaches—Greg Jackson and Mike Winkeljohn—demonstrated *Targeted Collection with Valid Analysis*. They knew Rousey was a judo Olympic medalist who specialized in takedowns and ground domination. Their task was to figure out how to mitigate that strength. They logged every step-in angle, every dropped left shoulder that preceded a lunge. They ran hundreds of rounds against sparring partners told to mimic Rousey's linear charge, drilling lateral exits, southpaw jabs to intercept the entry, and a check-hook to pivot off the cage. This was deliberate collection, disciplined analysis, and rehearsed counters—knowing the adversary's habits in detail and preparing for them until the response was automatic.

But superior collection was only half the equation. *Denial of Adversarial Intelligence Collection* was in play from the start. Holm gave vanilla interviews, posted generic workout clips, and generally avoided giving away any insights into her game plan. Media, starstruck by Rousey, failed to adequately spotlight Holm's pedigree—an 18-time world boxing champion with 100-meter-dash conditioning. What Rousey's camp could have learned, they never saw.

Fight night delivered the payoff. Rousey charged as the analysis predicted; Holm angled away. Rousey swung as the analysis predicted; Holm's head was gone. Commentators expected an early clinch; instead they saw a matador schooling a bull. Holm would strike on Rousey's approach and promptly back away, avoiding the takedown. When the fight did go to the mat, Holm would get up and back away, starting the cycle anew. As a result, Rousey took punch after punch. By Round 1's final horn, Rousey's nose bled, her mouth gaped for air, her eyes chased ghosts. Jackson's corner advice was a single line: "She's going to get a little more desperate this time." Rousey's corner offered platitudes, evidence of their own blind spots.

59 seconds into Round 2, Holm's left shin found Rousey's jaw. The champion collapsed face-first, consciousness trailing half a second behind. Spectators watched in shock as the chain of intelligence efforts unfolded in slow motion before their eyes: *intelligence collected, patterns catalogued, counters rehearsed, adversary intelligence collection denied, strategy executed.*

In a postfight interview, Jackson explained the win simply: "(Rousey) has been very successful doing the same things for a long time, and we were able to capitalize on that." Sun Tzu puts the counterintelligence lesson here eloquently: "Do not repeat the tactics which have gained you one victory, but let your methods be regulated by the infinite variety of circumstances."

Holm didn't win by outgunning the brawler—she won by seeing her more clearly while preventing the same in return. When both aims are executed in full, a perfectly timed jab can outweigh any uppercut of brute force.

### Negative Example — Yom Kippur Intelligence Failure

Seven years after the Six-Day War, Israel's intelligence apparatus appeared unrivaled. Its Unit 8200 intercepted Arab radio chatter; Mossad sources roamed Cairo and Damascus; Air Force overflights photographed every sandbag along the Suez Canal. Confidence hardened into dogma nicknamed *ha-Konseptsia*—"The Conception." It held that Egypt would not attack without air superiority, and Syria would not act without Egypt.

By September 1973, Egyptian President Anwar Sadat had decided to shatter the Conception, not by matching Israel's Air Force but by leveraging surprise, limited objectives, and dense SAM cover. His military simulated annual exercises near the Suez, training mobilization patterns until Israeli analysts filed them as routine. Soviet advisers quietly exited on October 5—an indicator dismissed as Kremlin politics. On October 6, Yom Kippur, sirens wailed.

Egyptian forces breached the Bar-Lev Line with water cannons and man-portable missiles, pushing deep before Israeli reserves could mobilize. Simultaneously, Syrian armor swept across the Golan Heights. The Israeli military was shattered in the first forty-eight hours, suffering hundreds of soldiers killed and the near-total destruction of its armored brigades on both the Suez and Golan fronts. Only hasty mobilization and a U.S. airlift turned the tide two weeks later.

Postwar commissions exposed failures in both aims. *Targeted Collection with Valid Analysis* broke down under the weight of assumption. Intercepts warning of unusual Arab logistics were downplayed. A Mossad source inside Egypt—coded "The Angel"—reported imminent attack, but analysts softened his language to "low probability." Two junior officers at Aman (Israeli military intelligence) flagged bridging-equipment counts as invasion indicators; their memos stalled in middle management. Critical fragments were collected but not correctly interpreted, falling victim to confirmation bias and overconfidence in the Conception.

*Denial of Adversarial Intelligence Collection* was equally lacking. Israel's strategic assumptions were so well known that Sadat could plan around them. Sadat's limited war aims, protected by dense air defenses, avoided the very conditions Israel believed were a prerequisite for any attack. No concerted effort was made to disguise readiness levels, introduce false patterns, or keep Egypt and Syria uncertain about Israeli thresholds. When your mental model becomes public knowledge, an adversary can design around it with devastating effect.

The Yom Kippur War remains a textbook case of what happens when both aims fail: vital data is misread or ignored, and the enemy's understanding of you is left unchallenged. In slow motion, the chain of intelligence failures is visible: *critical intelligence missed, patterns misinterpreted, counters absent, adversary intelligence collection unopposed, strategy exploited.* The result was strategic surprise, thousands of lives lost, and a diplomatic realignment that still shapes the region. In this campaign, the handmaiden of intelligence had been gagged, and the mistress of capability bled.

### Principles In Action — Reflection Questions

1. Are we executing *Targeted Collection with Valid Analysis* in a way that is thorough, relevant, and timely enough to detect both gradual drift and sudden shifts in adversary behavior?
2. Where do our routine disclosures—product demos, exercise schedules, hiring sprees—aid the adversary's collection, and how are we red-teaming these vulnerabilities to strengthen *Denial of Adversarial Intelligence Collection*?
3. Which analytic assumptions are treated as gospel, and when was the last time a formal dissent channel stress-tested them to ensure analysis remains valid?
4. In our counterintelligence posture, do we rely solely on *denial*, or do we also employ *deception* where appropriate, actively shaping what the adversary believes instead of just hoping they see less?
5. Who owns the continuous fusion of collection, analysis, prediction, and counterintelligence across all levels, and is that engine resourced for 24/7 cycles?

### Sources

- Bar-Joseph, Uri. 2005. *The Watchman Fell Asleep: The*

*Surprise of Yom Kippur and Its Sources*. Albany: State
University of New York Press.

- Heuer, Richards J. 1999. *Psychology of Intelligence Analysis*.
  Washington, DC: Central Intelligence Agency.
- Holm, Holly, and Ronda Rousey. 2015. Post-fight press
  conference transcript, UFC 193.
- Jackson, Greg. 2016. Interview on *The MMA Hour*.
- Keegan, John. 2004. *Intelligence in War: Knowledge of the
  Enemy from Napoleon to Al-Qaeda*. New York: Vintage
  Books.
- Okamoto, Brett. 2015. "Holly Holm's Coaches Thought
  Ronda Rousey Would Be Aggressor at UFC 193." *ESPN*,
  November 15, 2015. http://www.espn.com/mma/story/_/id/
  14134847/holly-holm-coaches-thought-ronda-rousey-
  aggressor-ufc-193
- Rabinovich, Abraham. 2004. *The Yom Kippur War: The Epic
  Encounter That Transformed the Middle East*. New York:
  Schocken.
- Sheridan, Greg. 2015. "Ronda Rousey Fight Results and
  Statistics." *ESPN*. http://www.espn.com/mma/fighter/_/id/
  2489290/ronda-rousey
- Sun Tzu. 1963. *The Art of War*. Translated by Samuel B.
  Griffith. Oxford: Oxford University Press.
- Ultimate Fighting Championship. 2015. *UFC 193: Ronda
  Rousey vs. Holly Holm* [Video]. YouTube, uploaded by UFC,
  November 15, 2015. Video, 3:17:30. http://www.youtube.
  com/watch?v=3j7WaTd8bkQ
- Winkeljohn, Mike. 2016. Seminar notes, Jackson-Wink
  Academy.

# PRINCIPLE 5
## ALIGN INTERNALLY

*"A house divided against itself cannot stand."* — Abraham Lincoln

I n any contest—tank battle, hostile takeover, zero-day hunt—the team that moves faster through the *see → decide → act* cycle wins. Alignment is the grease inside that cycle. If scouts spot an opening but logistics won't refuel on time, the gap closes. If cyber analysts flag lateral movement but compliance blocks emergency patches, attackers burrow deeper. A worthy rival can exploit your internal drift to strike first or strike cheaper. As we will see in Principle 7, Clausewitz warns that even simple plans warp once fighting starts; alignment keeps the warp from snapping the frame.

*Purpose* is your war aim. Without a shared aim, divisions optimize for private victories that add up to public defeat. This overlaps somewhat thematically with Principle 2, but it's necessarily more granular than simply defining victory conditions. In 1940 the French Army and Air Force both pledged to defend France, yet disagreed on whether the decisive fight would unfold on the ground or in the sky. Germany exploited that split, driving armor through the Ardennes while bombers punched holes in French command nets. Modern companies face the same risk: a sales unit chasing volume can over-promise features the product team lacks; a DevSecOps group demanding zero defects can freeze releases until market share is gone. As we examined in Principle 2, a single statement—liberate Kuwait, dominate streaming, protect customer privacy—focuses effort like a lens concentrates light. Planning (covered in Principle 7) provides the other supporting pillar. Internal alignment must be guided by both.

*Objectives* translate purpose into targets that weapons or code commits can hit. Good objectives honor the tactical clock. U.S. Marines call this "commander's intent": capture Hill 482 by dawn, not "improve security posture." Remember the "SMART" acronym? When SolarWinds patched Orion in 2020, its vague objective— "address customer feedback"—masked the urgency to close a supply-chain back door. Attackers had months to roam. Clear objectives compress the enemy's window.

*Incentives* fuel objectives. During World War II, the U.S. government awarded Army-Navy "E for Excellence" flags to factories that met exceptional production targets; the award became a powerful incentive, driving competition as entire communities took pride in displaying the flag and workers wore special pins. Contrast Wells Fargo's cross-sell quotas: employees opened fake accounts to satisfy a metric misaligned with genuine value-creation. In war, a misplaced metric costs lives; in business, it invites class-action lawsuits. Both drain capacity the moment real combat arrives.

*Routines* turn incentives into muscle memory—briefing rituals, code-review checklists, live-fire drills. The Israeli Defense Forces rehearse casualty evacuation in peacetime so that when a building collapses in Gaza, medics flow without orders. In cybersecurity, the "tabletop triage" ritual—who speaks to the board, who checks DNS logs—keeps incident response below the news-cycle threshold. Where routines differ, latency blooms. A marketing team on quarterly cycles cannot react to a sudden tweetstorm in minutes; a patching team bound to monthly windows cannot block a worm that spreads in hours.

∾

ALIGNMENT MUST SURVIVE TWO ENEMIES: *ENTROPY* AND *SHOCK*. ENTROPY is slow drift: key leaders leave, and laminated org charts freeze obsolete paths of authority. Shock is sudden—an earnings miss, a missile launch, a leaked exploit. Only habits built under lower stakes survive the spike. Toyota's Andon cord stops the line when a bolt won't seat, training reflexes that matter later when a supplier ships bad steel. The U.S. submarine force's two-minute "breathing report" at each watch change—intent, risk, next moves—proved critical during the April 2018 Syria strikes, when USS *John Warner* fired Tomahawk missiles: no officer wasted seconds asking, "What are we doing?"

Alignment is visible in freedom of maneuver. The German blitzkrieg owed as much to *Auftragstaktik*—mission-type orders—as to Panzers. Field captains knew the objective and could improvise routes when radios jammed. Modern analogues appear in Amazon's "two-pizza team" rule: small groups own end-to-end metrics, pivoting without chain-of-command bottlenecks. Both cases show alignment not as lock-step conformity but as shared intent enabling local initiative.

Tools help but do not govern. OKR dashboards, JIRA boards, and Slack huddles highlight drift early, yet cannot decide which arrow matters when two arrows diverge. That decision belongs to leadership, spoken aloud, often in tension. In 2007 Apple engineers argued whether to demo an Edge network or Wi-Fi prototype of the iPhone. Steve Jobs settled it decisively: with only six months to ship and EDGE available everywhere, he chose EDGE over Wi-Fi. Objective set, routines aligned, resources flowed. A competing handset maker spent another year debating and shipped late.

Misalignment often hides in day-to-day scheduling. A cybersecurity team may beg for a patch window while sales promises 99.99% uptime to close a deal. Both are acting in good faith, but without coordination, one team's success creates another's failure.

It also hides in language. "Secure" to an auditor might mean zero unauthorized access; to a product manager, it might mean "good enough to ship without legal risk." If these definitions aren't reconciled early, they tend to surface under maximum pressure, when misunderstandings are most costly.

Cost curves reveal the stakes. A McKinsey study reviewing large IT projects found that, on average, they run 45% over their original budgets while delivering 56% less value than forecasted. In kinetic war, the curve is steeper. During the 2003 invasion of Iraq, communications bottlenecks slowed the transmission of battlefield data; hours turned to days, and Baghdad's airport fell later than planned. A single senior-leader decision to reallocate bandwidth could have fixed it— but without alignment, the call never came.

Alignment also determines how fast you can learn mid-fight. Colonel John Boyd's OODA loop—observe, orient, decide, act— breaks if the "orient" picture is cluttered by conflicting incentives. A ragged loop cedes tempo to the adversary. Cyber red teams that report to the CISO, not to the dev lead whose code they break, provide honest orientation; bury them under product managers and the loop stalls.

Finally, alignment embodies trust. Soldiers follow orders into fire because they believe headquarters sees the same map and values the same outcome. Employees patch servers at 3 a.m. because they trust leadership to credit the save. Trust shortens deliberation; its absence forces endless checklists, slowing the fight.

The positive and negative cases ahead make the contrast concrete. Standard Oil forged alignment so tight that even its court-ordered breakup could not dissolve its operating DNA. AOL - Time Warner glued two giants together on paper but never aligned purpose, objectives, incentives, or routines, turning the biggest merger in history into the biggest write-off.

## Positive Example — Standard Oil's Integrated Machine

John D. Rockefeller began in 1863 with one Cleveland refinery and a simple *purpose*: deliver kerosene that was both cheaper and purer than anyone else's. Most refiners lost nearly half their crude to waste; Rockefeller's plant, run with church-ledger precision, lost less than

5%. When he formed Standard Oil in 1870, he carried that exacting culture into every refinery he acquired.

From that purpose, a single clear *objective* emerged: reduce the cost per gallon. Transportation was the largest expense, so Rockefeller negotiated fixed-rate contracts with railroads — controversial but legal at the time — and later built his own pipelines to bypass rail entirely. Pipeline managers were told, in writing, that every freight dollar saved was a dollar that could lower the retail price and keep competitors out. By 1870 Standard controlled 4% of U.S. refining; by 1879 it controlled 90%. Kerosene prices fell from twenty-six cents a gallon in the Civil War era to eight cents by 1885, even as Rockefeller paid record dividends.

*Incentives* reinforced the objectives. Plant superintendents who cut waste received a share of the savings. A foreman who saved $5,000 in solvent might see half that amount as a bonus. Because bonuses were tied to verified cost drops, gaming the system was nearly impossible — another reason uniform ledgers mattered. One foreman famously trimmed solder on tin cans by a fraction of an ounce, saving tens of thousands of dollars yearly; Rockefeller called him "a partner in the result."

Daily *routines* kept alignment tight. Every refinery used the same morning checklist — boiler pressure, reflux ratios, waste-water clarity. If a metric drifted, the local superintendent telegraphed headquarters before lunch. Standard also rotated managers between plants, spreading best practices and breaking down local fiefdoms. A Baltimore chemist might spend six months in Ohio, then take an accounting post in New York, gaining multiple perspectives and carrying best practices with him.

Critics hammered Standard for crushing independent refiners, and in 1911 the Supreme Court dissolved the trust. Yet even the breakup acknowledged the integrated power of the machine. Of the thirty-four spin-off companies, many — Exxon, Mobil, Chevron — remained dominant because they had inherited the same alignment DNA: shared *purpose*, cascading *objectives*, audit-backed *incentives*, and *routines* that turned monthly variance into morning action. Standard

Oil became the classic MBA case study for alignment done right —
proof that tight internal links can drive long-term external
dominance.

### Negative Example — AOL and Time Warner: The Merger That Never Merged

On January 10, 2000, AOL CEO Steve Case and Time Warner CEO
Gerald Levin announced the largest merger in U.S. history — a $165
billion deal meant to fuse "old media's content" with "new media's
pipes." The plan was simple on paper: combine AOL's massive online
audience with Time Warner's deep library of television, film, and
publishing assets. The ceremony oozed synergy. Wall Street
applauded, the press gushed, and employees on both coasts
wondered how two corporate cultures would mesh. They soon
found out.

*Purpose* diverged from the start. AOL's goal was to keep its dial-up
subscribers inside its own proprietary chat rooms and content areas,
rather than sending them to the open internet. Time Warner's goal
was to produce HBO dramas, feature films, and magazines for the
widest possible distribution. No single unifying statement reconciled
those aims. Engineering meetings devolved into arguments about
how restrictive to make digital rights management (DRM). Marketing
teams clashed over whether to sell music and movie CDs by mail or
stream video clips online.

*Objectives* conflicted as well. AOL measured success in new
subscriber numbers. Time Warner measured success in box-office
revenue, television ratings, and magazine ad sales. AOL product
managers pushed for pop-up ads across Time Warner's web proper-
ties; Time Warner editors resisted, arguing it would alienate readers
and viewers.

*Incentives* widened the rift. AOL executives held stock options that
had surged in value during the dot-com boom; Time Warner executives relied on pensions and steady dividend stocks. When Yahoo's
share price sagged, AOL managers chased short-term ad clicks to

protect their payouts, while Time Warner journalists argued for the patience needed to nurture longform journalism. Neither side trusted the other's pay structure or motives.

*Routines* sealed the misalignment. AOL's teams were accustomed to pushing software updates monthly; Time Warner's magazine teams worked on sixty-day production cycles, and its television units on even longer schedules. Conference calls overflowed with acronyms each side barely understood. One senior editor quipped that meetings felt like "a blind date chaperoned by lawyers." Integration teams produced slide decks but never established a common workflow.

When the dot-com bubble burst in late 2000, AOL's dial-up subscriptions stalled, and the merger's much-touted synergies became harder to justify. By 2002, the company took a $99 billion goodwill write-off — the largest in history at the time. Levin resigned, followed by Case. Time Warner gradually shed the AOL name, eventually spinning it off in 2009 at a fraction of the original value. Shareholders who held through the deal lost about 80% of their investment.

Alignment failed at every ring: the two companies never shared a unified *purpose*, their *objectives* pulled in opposite directions, *incentives* rewarded contradictory behaviors, and their *routines* clashed at the operational level. No dashboard, re-org, or culture workshop could bridge the structural gulf. The result remains one of the most-cited cautionary tales in corporate history — a merger that never truly merged.

## Principles in Action — Reflection Questions

1. Is our *purpose* stated in a single, unifying sentence that every team and department can articulate the same way?
2. Are our *objectives* specific, time-bound, and clearly linked to that purpose, with no cross-departmental conflicts?

3. Do our *incentives* reinforce those objectives, and are they structured so that gaming one metric can't damage the others?
4. Are our *routines*—meetings, briefings, reviews, drills—aligned with both the objectives and the tactical tempo we face?
5. Taken together, do our purpose, objectives, incentives, and routines trace a single line from top floor to shop floor, even under shock or entropy?

**Sources**

- Chernow, Ron. 1998. *Titan: The Life of John D. Rockefeller, Sr.* New York: Random House.
- Cybersecurity and Infrastructure Security Agency. 2020. "Alert AA20-352A: Mitigations for the SolarWinds Orion Supply-Chain Compromise." December 2020.
- Isaacson, Walter. 2011. *Steve Jobs.* New York: Simon & Schuster.
- Jones, Arthur W. n.d. Quoted phrase popular in organizational management studies.
- Lincoln, Abraham. 1953. *The Collected Works of Abraham Lincoln,* vol. 2. Edited by Roy P. Basler. New Brunswick, NJ: Rutgers University Press.
- McKinsey & Company. 2012. "Delivering Large-Scale IT Projects on Time, on Budget, and on Value." August 2012.
- Munk, Nina. 2004. *Fools Rush In: Steve Case, Jerry Levin, and the Unmaking of AOL Time Warner.* New York: Harper Business.
- Ohno, Taiichi. 1988. *Toyota Production System: Beyond Large-Scale Production.* New York: Productivity Press.
- Office of the Comptroller of the Currency. 2016. "Consent Order #2016-113, Wells Fargo Bank, N.A." September 8, 2016.

- USNI News. 2018. "USS John Warner Launches Tomahawks in Syria Strikes." April 14, 2018. http://news. usni.org/2018/04/14/uss-john-warner-launches-tomahawks-in-syria-strikes
- Womack, James P., Daniel T. Jones, and Daniel Roos. 1990. *The Machine That Changed the World.* New York: Free Press.
- Yergin, Daniel. 1991. *The Prize: The Epic Quest for Oil, Money, and Power.* New York: Simon & Schuster.

# PRINCIPLE 6
## ALLY WITH PURPOSE

*"If you want to go fast, go alone. If you want to go far, go together."*
— Traditional Proverb

lliances sound obvious—of course more hands can lift a heavier load—yet history is littered with coalitions that stalled, splintered, or even crippled the partners they were meant to empower. The problem is not the instinct to join forces; it is the failure to ally with a *clear, shared purpose*. When actors merge strength without aligning aims, logistics, and trust, they entangle rather than multiply. I have sat at tables where partnerships saved months of toil—and at others where half-translated memos turned potential allies into resentful spectators. The difference lay in whether the coalition satisfied five conditions I came to call the *Five Cs:*

1. **Complementarity** – Each partner supplies a capability the other lacks, avoiding redundant muscle.
2. **Common Purpose** – Participants want the *same* end state, not just the defeat of a mutual foe. (The enemy of my enemy is *not necessarily* my friend, as we'll explore below).

3. **Commitment Mechanism** – Treaties, equity, or doctrine make defection costly and cooperation reflexive.
4. **Continuous Communication** – Shared language, compatible tools, regular rehearsal turn theory into real-time coordination.
5. **Calibration & Adaptation** – The alliance revisits goals and roles as the environment shifts, preventing yesterday's synergy from becoming tomorrow's drag.

Remove any one C and cracks spread. Complementarity without common purpose breeds tension; common purpose without commitment tempts freeloading; commitment without communication paralyzes the team; communication without calibration devolves into endless meetings. Get all five right and power compounds like interest.

Ever since Thucydides chronicled Sparta and Athens, alliances have pivoted on threat perception. Modern strategists echo the same sentiment: the enemy of my enemy is my friend. Shared fear makes weak glue. Fear recedes, new threats arise, and partners drift. Purpose endures. When Apple partnered with Intel to use its processors, the purpose was to smoothly migrate the Mac to an x86 architecture; once complete, the alliance dissolved amicably. Conversely, early WWI Britain and France shared dread of Germany but diverged on nearly everything else; their Entente stumbled under contradictory peace visions and mismatched command: a story we will revisit.

### *Complementarity*: The Physics of Mutual Benefit

The physics metaphor is apt: vector addition only produces a larger arrow if angles align; combine force at 180 degrees and net movement is zero. Allies must deliver orthogonal strengths—speed paired with stealth, cash with creativity, territory with technology. The coral reef team of grouper and giant moray eel—open water sprinter meets tunnel specialist—is nature's elevator-pitch for complementarity. Human equivalents abound: semiconductor designers partnering with Taiwan's TSMC fabs; NATO pairing U.S. strategic airlift with European rapid-reaction infantry; humanitarian NGOs matching medical know-how to local knowledge.

### *Common Purpose*: More Than a Mutual Enemy

Shared fear may rally partners to the same banner, but only a shared vision of victory ensures they don't turn on each other once the fighting stops. The most common strategic error is confusing a mutual adversary with a common goal. An alliance built on pure opposition is inherently unstable; it is defined by what it seeks to destroy, not what it intends to build. True purpose must be articulated as a positive, concrete end state. The alliance between the Western powers and the Soviet Union in World War II was united only by the goal of defeating Nazi Germany; the moment that enemy fell, the alliance predictably fractured into the Cold War. Conversely, the success of the Apollo Program stemmed from a purpose defined not as 'beating the Soviets,' but as achieving a specific outcome: landing a man on the Moon and returning him safely. That level of precision is the only true glue for a coalition under stress.

### *Commitment*: Bonds That Bite

Commitment without teeth invites abandonment at inconvenient hours. Roman historian Livy quipped that treaties break when fear subsides. Modern commitment mechanisms convert rhetoric into deterrent cost: cross-licensing IP, escrow accounts, equity swaps, or mutual-defense clauses. Even social-capital pledges—public declarations broadcast to stakeholders—raise reputational stakes. Good commitment design balances flexibility and glue: too loose, and partners bolt; too tight, and they feel trapped.

## *Communication*: Pulse of Coalition Life

The best complementarity is useless if partners cannot synchronize. Language barriers, data-format mismatches, incompatible doctrine—these frictions turn would-be synergy into attrition. Alliance architects therefore invest in translation layers: NATO brevity codes, SWIFT financial messaging, Joint All-Domain Command and Control standards. Training exercises, not press releases, reveal whether routers talk and rifles fire on the same count.

## *Calibration*: Alliances Are Living Systems

Goals drift, resources fluctuate, enemies learn. Coalitions that lock doctrine in amber become self-parodies. The Five Cs therefore close with calibration—periodic review of threat, purpose, and payoff. Post-OPEC embargo, Shell rewrote scenario partnerships every two years; the U.S.–Japan defense pact has issued nine "Guidelines for Cooperation" updates since 1951; even ant colonies swap worker castes in response to rainfall. An alliance that cannot adapt commits slow suicide.

## Ethical Undercurrent

Alliances amplify power and therefore responsibility. Exploitative coalitions—using small partners as expendable buffers, extracting data without returns—erode legitimacy and seed defection. Principle 28's moral compass points here: steward allies as co-architects, not disposable cogs.

~

WITH THESE FRAMEWORKS IN PLACE, WE CAN APPRECIATE TWO narratives: one where purpose-driven complementarity transforms improbable partners into lethal collaborators, and one where vague purpose and lax mechanics nearly doom a continental coalition before it matures.

**Positive Example — Giant Moray & Grouper: Reef-Floor Alliance**

Beneath the Red Sea's kaleidoscopic corals, the giant moray eel winds through crevices like living rope. It excels at ambush but struggles in open pursuit. The Red Sea grouper, by contrast, is a powerful missile —deadly in bursts across open water yet too bulky for labyrinthine reef mazes. Individually, each predator posts prey-capture success rates barely above 5%.

Marine biologists, led by Redouan Bshary, once shadowed these fish with waterproof laptops and waterproof patience. They recorded a ritual: grouper spots prey wedged beyond its girth, then flips vertical and performs a "headstand–shudder" signal—mouth agape, pectoral fins quivering. A moray eel, noticing the dance, glides from a nearby crack and invades the hole. Chaos ensues. If the prey bolts into blue water, the grouper accelerates to devour it. If it dives deeper, the eel coils and clamps. Scientists logged a remarkable statistic: paired hunts succeeded in over 50% of trials—roughly a tenfold jump versus solo attempts.

Why does this alliance glow as a case study? Because it embodies every C:

- **Complementarity** – Open-water acceleration blends with tunnel agility.
- **Common Purpose** – Shared caloric gain; the meal is split by immediate devouring, leaving little room for posthunt betrayal.
- **Commitment Mechanism** – Repetition and direct payoff; a moray that eats the grouper now will be hungry later; defection today forfeits future meals.
- **Communication** – Simple, high-contrast gestures understood across species lines.

- **Calibration** – Pairings recur only when prey type
  demands it; each fish exits alliance when goals diverge.

For human strategists, the parable is vivid. Replace reef with
supply chain: airlift is handled by Atlas Air; last-mile delivery by
Amazon's ground network. Fuse both under peak-season surge, and
parcels flow like fish fleeing an eel. In cyberspace, incident-response
teams (grouper) coordinate with gray-hat vulnerability researchers
(moray) to flush exploits into monitored sandboxes. Complementarity, purpose, commitment, communication, calibration—the blueprint works above and below sea level.


## Negative Example — Entente Cordiale Under Fire

On April 8, 1904, Britain and France buried centuries of rivalry in a
colonial handshake dubbed the Entente Cordiale. It settled legacy
Nile basin disputes and divided West-African spheres. What it did
not settle was how to fight together if Germany struck. No joint
command, no standardized rail plan, no artillery caliber alignment.
Politicians assumed goodwill would bridge details; generals prayed
their opposite numbers spoke the same staff-college dialect.

August 1914 shredded those hopes. Germany's opening plan sent
its right wing—the 1st–3rd Armies, with elements of the 4th—
through Belgium; the 4th–5th fought in the Ardennes, while the 6th–
7th attacked in Alsace-Lorraine. France mobilized under Plan XVII—
an all-out dash toward Alsace-Lorraine. Britain's tiny but professional
BEF landed at Boulogne with just six infantry divisions and a
commander, Sir John French, whose name foretold confusion: he
distrusted his French allies, and they returned the sentiment.

The Five Cs cracked in sequence:

- **Complementarity** existed—French mass conscripts offset
  British rifles—but mismatched doctrine negated synergy.
- **Common Purpose** was murky. France dreamed of
  reclaimed provinces; Britain feared naval blockade.

- **Commitment** was thin: no treaty required Britain to stay if losses ballooned.
- **Communication** failed. Telegraph lines were cut, liaison officers lacked shared maps, artillery grids used different meridians.
- **Calibration** vanished. French staff clung to offensive à outrance even after machine-gun slaughter; BEF generals demanded defensive anchors.

At Mons (August 23) the BEF fought a delaying action then retreated; French high command misread the move as British rout. During the eleven-day "Great Retreat," British columns and French armies drifted fifty miles apart, leaving German cavalry to probe gaps. Supplies jammed at rail choke points; secrecy obsessed each staff so deeply that they hid plans even from partners. When the Marne counterstroke finally halted Germany in September, historians called it a miracle. In truth, it was logistics luck: Paris taxis shuttled reserve regiments, and a German gap opened between two armies neither ally anticipated.

Only in 1915 did the creation of the inter-allied Chantilly conferences, followed by a Supreme War Council in 1917, begin repairing commitment and communication layers. But the price of early misalignment was staggering: hundreds of thousands of casualties, lost initiative, and trench warfare that froze the front for four grinding years.

The Entente's stumble proves that alliances formed around vague fear degrade under adaptive pressure. Had Britain and France scripted complementary operations, codified railway allocations, exchanged artillery doctrines, and rehearsed calibrations, the Marne might have been a planned pivot, not a blood-soaked improvisation.

## Principles In Action — Reflection Questions

1. Which ally supplies a capability we lack, and have we formalized the exchange so that both sides benefit without redundancy?
2. Have we articulated an identical, positive end state with each major partner, rather than relying on shared opposition to a common rival?
3. What binding mechanism—legal, financial, operational, or reputational—ensures that our partners remain committed under stress?
4. Do our communication channels, formats, codes, and rehearsal schedules function seamlessly in live conditions, not just in planning documents?
5. When was the last time we reviewed and adjusted roles, responsibilities, and rewards with our allies to account for changes in threats, technology, or resources?

## Sources

- Apple. 2005. "Apple to Use Intel Microprocessors Beginning in 2006." News release, June 6, 2005. http://www.apple.com/newsroom/2005/06/06Apple-to-Use-Intel-Microprocessors-Beginning-in-2006/
- Axelrod, Robert. 1984. *The Evolution of Cooperation.* New York: Basic Books.
- Bshary, Redouan, Andrea Hohner, Karim Ait-el-Djoudi, and Hans Fricke. 2006. "Interspecific Communicative and Coordinated Hunting between Groupers and Giant Moray Eels in the Red Sea." *PLOS Biology* 4, no. 12: e431. https://doi.org/10.1371/journal.pbio.0040431
- Clayton, Tim, and Phil Craig. 2000. *Finest Hour: The Battle of Britain.* London: Hodder & Stoughton.

- Keegan, John. 1998. *The First World War.* New York: Vintage Books.
- Livy. 1919. *Ab Urbe Condita,* Bk. I. Translated by B. O. Foster. Cambridge, MA: Harvard University Press.
- NATO. n.d. *Founding Treaty Texts.* North Atlantic Treaty Organization.
- Tuchman, Barbara W. 1962. *The Guns of August.* New York: Macmillan.
- Wainwright, C. Martin. 2004. "Entente Cordiale in Cartoon and Caricature." *History Today* 54, no. 4: 42–48.
- Whitby, Andrew. 2020. "If You Want to Go Fast ..." Blog post, December 25, 2020. http://andrewwhitby.com/2020/12/25/if-you-want-to-go-fast/

# PRINCIPLE 7

## PLAN FOR ALL SCENARIOS

*"In preparing for battle I have always found that plans are useless, but planning is indispensable."* — Dwight D. Eisenhower

I first learned the tyranny of contingency in a dusty classroom at the U.S. Army Intelligence School at Fort Huachuca. A drill sergeant—rough voice with humor to match—slapped a marker across the whiteboard: PACE. *Primary, Alternate, Contingency, Emergency.* "Memorize it," he barked, "because no plan survives the first five minutes of battle." The line echoed Helmuth von Moltke's nineteenth-century dictum—*no plan survives first contact with the enemy*—later paraphrased by every commander who watched a morning brief's tidy plan evaporate before lunch. Nearly thirty years in intelligence have since proved the sergeant right: a beautiful plan is a short-lived luxury. But *a mind honed by rigorous planning* is an enduring asset.

Why does reality shred blueprints? Three forces conspire.

    1. **Complexity.** Even a mid-sized adversarial system—an infantry company, a hedge-fund portfolio, a ransomware gang—contains thousands of variables. Interactions

between variables explode combinatorially. The precise
state of that system tomorrow occupies one slot in a
cosmic lottery of possibilities.

2. **Adaptation.** An intelligent opponent does not sit still
while you execute step three; they push back, reroute,
innovate, deceive, panic, surprise. Their moves shift the
terrain under your boots.

3. **Noise.** Weather rolls in, supply trucks break axles, market
rumors spike volatility, an analyst mistypes a subnet
mask. Randomness fogs vision and jams timing.

Eisenhower and Moltke weren't saying *don't plan.* They were
saying: don't worship the plan. What matters is *planning culture*—the
habit of thinking through multiple futures, rehearsing pivots, and
wiring in backup options before chaos arrives. Done well, it makes
disorder feel less like a shock and more like a drill gone live.

## Scenario Breadth, Depth, and Layered Response

Three traits separate real planning from wishful thinking.

First, *breadth.* Imagine futures that break your straight-line forecast. What if capital dries up? What if a key supplier folds? What if
the worst day hits on a public holiday? Teams that only picture business as usual are the ones most easily blindsided.

Second, *depth.* Don't just think about the first-order effect. A 30%
tariff isn't just a line-item cost—it can shift consumer loyalty, trigger
currency swings, and invite retaliation from regulators. Ripples
become waves.

Third, *layered response.* Align authorities, budgets, and communication channels for each level of disruption. That's what the Army's
PACE system teaches: radio, satellite, courier, each ready if the one
before fails. A software company might line up primary cloud
servers, a failover region, an on-premises cluster, and a cold backup
tape.

The details differ, but the pattern repeats. Cyber defenders write playbooks for stolen credentials, lateral movement, mass encryption, misinformation blowback. African elephants keep multiple fallback routes for droughts. A biotech startup lines up alternate reagent suppliers in case geopolitics cuts off trade. Across domains, planning culture creates room to maneuver when reality shreds the script.

## Planning as Cognitive Weight Training

Planning also sharpens perception. When teams storyboard dozens of futures, they seed their working memory with pattern libraries. Later, when reality delivers something similar, recognition is faster. Decisions come quicker. In OODA-loop terms, breadth training speeds up *Observe–Orient.* Depth training smooths *Decide–Act.*

## PACE—The Memory Hook

The Army's PACE ladder endures because it makes leaders show their work. Ask a section chief, *What's our alternate comms channel?* If the answer takes more than five seconds, the plan is weak. PACE drags hidden gaps into daylight.

The same logic applies beyond comms. What's your primary market? Your backup revenue stream? Your contingency line of capital? Your emergency asset sale? It's uncomfortable to ask those questions—but far better in a boardroom than on live television under floodlights.

## Imagination versus Inertia

Humans under-imagine unlikely futures. Psychologists call it *normalcy bias.* We picture tomorrow as an extension of today because it feels safe to the story-telling brain. Scenario planning pushes back. It asks us to explore futures that seem awkward, expensive, or culturally out of character.

Pierre Wack, the man who introduced scenario thinking to Shell, called it the *"gentle art of reperceiving."* The goal isn't to guess perfectly.

It's to stretch the retina of the organization, so it sees more than one line on the horizon.

## Dependencies, Tail Risk, and Black Swans

Every system hides a weak link. Maybe it's one engineer who knows the only password. Maybe it's a factory that makes a single pigment your product needs. Maybe it's a machine-learning model dependent on an outside company's API. Everything looks steady—until that one piece disappears.

The rare disasters are the ones that matter most. A once-in-a-century flood. A market crash no one predicted. A virus that leaps across continents. These are *tail risks*—low odds, but catastrophic impact. They rarely show up in forecasts, but they can ruin you if you don't prepare.

That's where the phrase *"black swan"* comes in. Europeans once assumed all swans were white. Then Dutch explorers reached Australia in 1697 and found black ones everywhere. Overnight, "impossible" became real. Nassim Nicholas Taleb later used the term to describe rare, high-impact shocks—like 9/11 or the 2008 financial crash—that seem obvious only in hindsight.

Planners who never imagine a black swan are stunned when it arrives. Planners who do imagine it—who ask, *"what if the unthinkable happens?"*—respond with calm. They already have an outline for what to do next.

## Emotional Steadiness Under Surprise

Teams that rehearse alternate futures carry themselves differently when the main path collapses. Shell managers in 1973 walked into the oil embargo with signed capital sheets ready for the North Sea. Nokia executives in 2007 froze as the iPhone devoured their roadmap. Calm doesn't come from temperament alone. It comes from preparation.

## Ethical Angle: Transparency versus Manipulation

Planning only works if people are free to voice bad news. If scenario teams fear punishment for "negative thinking," the whole exercise collapses. Shell protected its planners from political retaliation. Nokia drowned dissent in middle-management filters. Smothering uncomfortable futures is more than bad management—it's an ethical failure.

## Why Plan When Plans Fail?

Because planning transforms uncertainty into structure. You may never use Plan B word-for-word, but thinking it through clarifies which resources to stock, who needs training, and which decisions to predelegate. Fighter pilots brief alternate landing strips every morning. When an engine flames out, the checklist feels familiar. The battle was partly won on the runway, long before the dogfight over unknown terrain.

## Positive Example — Shell's 1970s Scenario Planning

We've mentioned Shell a couple of times. Let's dive in deeper. Royal Dutch / Shell in the mid-1960s was an oil major but not *the* oil major. Its reserves lagged Exxon's; its political clout trailed BP's. Enter Pierre Wack, a former economist with a taste for Sufi philosophy. He believed forecasting was hubris cloaked in decimals. Instead, he urged his Group Planning team to write *stories*—rich, internally coherent futures that captured political mood swings, cartel temptations, and technology jolts.

By 1972 Shell's planners had developed several narratives. One asked: *What if Arab producers weaponize oil prices to punch at Western allies?* They built the storyline down to ship turnaround times, European refinery specs, and consumer psychology under rationing. Executives, skeptical but intrigued, approved modest hedges: refinery retrofits for heavier crudes, acceleration of North-Sea seismic surveys, long-lead tanker contracts.

October 1973 proved the thought experiment prophetic. The Yom Kippur War erupted; OPEC embargoed nations supporting Israel. Spot prices quadrupled; tankers queued at Rotterdam. While competitors scrambled, Shell pulled scenario binders off shelves and executed. Refineries previously tooled for Arab Light could now crack Gulf Heavy; capital earmarked for North Sea rigs activated; shipping slots reserved at lower precrisis rates sailed under Shell flags. By 1974 Shell's financial position had strengthened markedly, and in the years that followed it gained retail share across Europe.

The cultural payoff outlasted the embargo. Scenario sessions became ritual—teams debated "tripolar currencies," "a carbon-constrained world," and "Asian demand explosion." When oil collapsed in 1986, Shell's balance sheet absorbed the shock with fewer layoffs, having rehearsed a price-crash storyline. Employees internalized the mantra: *No single view is sacred; the only sin is failing to imagine.* "A failure of imagination" is a familiar phrase for anyone from an intelligence background.

Wack later framed their success as "reperceiving" rather than predicting. By glimpsing multiple futures, Shell executives disentangled identity from any one forecast. Shareholders cared less whether one narrative came true; they cared that profits endured whichever narrative unfolded. Scenario breadth had converted uncertainty from fatal threat into navigable sea lanes.

### Negative Example — Nokia's Smartphone Collapse

By 2007 Nokia's name was synonymous with mobile. Its hardy phones dominated emerging markets; its Symbian OS owned nearly half of global smartphone share. Anyone old enough to have had a mobile phone in those early years can attest to how much Nokia dominated the market. If you had a mobile phone, chances are it was a Nokia. They were absolutely *entrenched.* The company ran elaborate product-roadmap meetings, but those meetings revolved around *a main plan*—iterative camera upgrades, hardware segmentation by price band, and incremental menu tweaks. I remember that calculator-like menu well.

Steve Jobs unveiled the iPhone in January 2007. Nokia's senior engineers flagged two disruptive vectors: all-screen interface and a developer-led ecosystem. One memo warned, "We are now competing with a computer company, not a handset maker." Yet in the next quarterly review, Symbian unit heads argued that Nokia's brand loyalty and distribution scale rendered Apple a niche threat. Alternate futures—touch UI saturates globally; operator subsidies fund expensive hardware; app stores erode operator walled gardens —failed to make the cut.

Engineers proposed advancing their internal Linux touch OS—a project based on its *Maemo* software—to hedge against the new threat. Finance declined the head-count request, citing focus on Symbian polish. R&D pitched a rapid-prototyping lab; leadership redirected funds to marketing campaigns promoting the N95's five-megapixel camera.

Meanwhile, Android opened its SDK in late 2007. HTC's G1 arrived in 2008, proving that an open-source ecosystem could iterate faster than proprietary firmware. Still, Nokia clung to its dominant-player math. Market-share projections extended current curves; worst-case slides shaved only a few points.

Then reality hit: By late 2010, Android and iOS combined controlled over 46% of the market; Nokia's Symbian had slid to 32%. Developers flocked to Apple's App Store, then Google Play. Consumers now cared less about megapixels, more about pinch-zoom and then-brand-new Instagram filters. Nokia's board hired Stephen Elop, who wrote the infamous "burning platform" memo: "We have poured gasoline on our own burning platform." Translation: we planned for a world that no longer existed.

Elop struck an alliance with Microsoft. Windows Phone arrived too late. Flagship Lumia devices earned praise but not volume; developers, already invested in two ecosystems, ignored a third. By 2013 Nokia's smartphone share hovered near 3%; Microsoft acquired the handset division for $7.2 billion—about one-tenth Nokia's 2007 valuation.

What doomed Nokia was not ignorance of touchscreen potential; it was failure to walk the PACE ladder. Primary plan: Symbian iteration. Alternate: delayed Maemo/Linux pivot. Contingency: hardware OEM option. Emergency: asset sale. Only the emergency rung survived, activated too late to rescue equity holders. Normalcy bias, dressed in brand swagger, smothered scenario breadth.

## Principles in Action — Reflection Questions

1. Have we mapped at least three futures that contradict our baseline forecast, and assigned owners to rehearse each?
2. When did we last test our PACE ladder end-to-end—with real systems, real people, and realistic time pressure?
3. Which single dependency could an adversary or accident remove tomorrow, and what prefunded pivot activates if it happens?
4. Do junior analysts feel psychologically safe to propose "ridiculous" scenarios, and do leaders track those proposals to closure?
5. If our flagship strategy vanished overnight, what alternate revenue, mission, or capability would keep us solvent for six months?

## Sources

- Elop, Stephen. 2011. "Burning Platform." Internal memo, Nokia Corporation, February 2011.
- Gartner. 2011. "Gartner Says Worldwide Smartphone Sales to End Users Increased 72 Percent in 2010." Press release, February 9, 2011.
- IDC. 2014. "Smartphone OS Market Share, 2013." *IDC Quarterly Mobile Phone Tracker,* February 2014.

- Jobs, Steve. 2007. "iPhone Launch Keynote." Speech, Macworld Conference & Expo, San Francisco, January 9, 2007.
- Moltke, Helmuth von. 1892. *Militärische Werke,* vol. 1. Berlin: Ernst Siegfried Mittler & Sohn.
- Nixon, Richard M. 1957. "Address at the National Defense Executive Reserve Conference." Speech, Washington, DC, November 14, 1957.
- Nokia Corporation. 2006. *Annual Report.* Espoo, Finland: Nokia Corporation.
- Shell Group Planning. 1983. *Scenarios 1973–1983.* London: Royal Dutch/Shell Group.
- Taleb, Nassim Nicholas. 2007. *The Black Swan: The Impact of the Highly Improbable.* New York: Random House.
- Wack, Pierre. 1985. "Scenarios: Uncharted Waters Ahead." *Harvard Business Review* 63, no. 5: 73–89.

# PRINCIPLE 8

## TEST AGAINST REALISTIC RESISTANCE

*"Prowess, unless it has an adversary, collapses ... we can only know its greatness and its power when it has shown by its endurance what it is capable of."* — Seneca the Younger

A new capability gleams on launch day: fighter pilots run perfect missions in the simulator, surgeons perform clean procedures on cadavers, and cyber teams watch their algorithms crush sample data. Then reality arrives: the enemy pilot doesn't fly by the script, a real patient's body reacts in messy, unpredictable ways, and a frustrated user clicks in places no test ever covered. The shine dulls fast. What separates systems that thrive from those that implode is the toughness of the tests they faced before the curtain rose. Seneca issued the warning two millennia ago, yet organizations still graduate those tested only against compliant opposition—and then wonder why the first punch lands deafeningly hard.

Realistic resistance means resistance that doesn't play along. It sidesteps your safest assumptions, lures you toward your strengths to expose hidden weaknesses, and changes shape between one encounter and the next. Think of it as a ladder of difficulty.

- At the bottom is *inert resistance*—like throwing punches at a heavy bag. It absorbs your strikes, but it never thinks, never moves unpredictably, never hits back.
- Next comes *compliant resistance*—a sparring partner who sticks to the script. They throw the punches you expect, in the order you expect them. Useful for drills, but no real surprise.
- Above that is *uncooperative resistance*—an opponent who genuinely tries to win, but still within the known rules of the match. They push back hard, but the boundaries are fixed.
- At the top is *adaptive resistance*—an adversary who breaks those boundaries. They change the rules as you fight, find ways to cheat mid-match, or flip the game board entirely.

Most training plateaus at *compliant resistance.* Why? Comfort, time pressure, legal caution, and the natural human aversion to embarrassment. It is exhilarating to demonstrate mastery when the environment plays nice. The problem surfaces only later, when real stakes arrive and the environment refuses to cooperate. All at once, that sense of mastery reveals itself as a mirage built on gentle feedback.

Intentional practitioners climb the resistance spectrum on purpose. They build independent red teams exempt from production deadlines and empowered to cheat. Netflix's Chaos Monkey randomly kills servers in live clusters to ensure resilience. SpaceX deliberately detonates test articles because each explosion feeds data to the next Starship iteration. AWS runs "GameDay" exercises that simulate real-world outages, where engineers break dependencies until dashboards light up; observers track mean time to detect (MTTD) and mean time to repair (MTTR) under stress.

Effective adversarial validation respects three design laws. First, independence: the team that builds cannot be the team that breaks, or confirmation bias creeps in. Second, authority: testers must have license to escalate, pivot, and invent surprises—as real opponents would—without begging permission in the heat of the moment. Third, rapid feedback: lessons must funnel to builders quickly, before memory erodes or defensive posturing sets in. Cultures that punish bad news strangle this feedback loop; no one will reveal the vulnerability that could imperil promotions.

(If you haven't noticed yet, the necessity of tolerating "bad news" as a strategic advantage is a recurring theme in this book, holding across many of the thirty principles).

Psychology matters. Adaptive tests inoculate nerves by simulating adrenal conditions. Brazilian Jiu-Jitsu students get outgrappled and "tap out" hundreds of times weekly, rewiring their fear response so that they relax under tight chokes and focus on escape. Pilots drilled in motion-based simulators encounter startle factors and spatial disorientation; those who only practiced instrument scans on desktop screens often freeze when the cockpit shakes. Stress grafted onto rehearsal becomes composure in crisis.

But realism obviously must be bounded by responsibility. A penetration test that shuts down a hospital crosses from validation into malpractice. Red-team operations—like combat sports—require guardrails: operational limits, abort codes, and postengagement debriefs. These aren't constraints on creativity but safeguards that keep experimentation tethered to purpose. Even judo was born as a safer offshoot of jujutsu, designed to allow full-resistance training without permanent harm. Brazilian Jiu-Jitsu follows that same lineage: structured for real intensity, minus real injury. The result of that structure—how resistance accelerates adaptation—will be explored momentarily.

This need for testing against realistic resistance transcends domains. Immune systems train by letting T-cells duel false self-markers in the thymus; only resilient cells graduate to defend the body. AI self-play systems like AlphaZero become superhuman by confronting versions of themselves learning at equal pace. Economies stress-test banks with fictional recessions to gauge capital adequacy. Where adaptive resistance shows up early, collapse shows up late. Where it is ignored, the timeline flips.

Of course, this preconflict validation is the ideal, but it isn't always possible. When a force must engage without extensive rehearsal, the principle adapts: seek resistance early, but on a small scale. Initial contact becomes a form of live, adaptive testing. Small, controlled skirmishes—like a military reconnaissance-in-force or a product's soft launch in a minor market—are not designed to win the war, but to learn how the enemy fights. They allow a force to absorb the first punch when the stakes are low, so that the decisive blow can be delivered with precision when the stakes are high.

Two modern stories illuminate the stakes. Brazilian Jiu-Jitsu conquered early mixed martial arts because its practitioners treated every practice session as unscripted combat against fully resisting bodies. Knight Capital lost half a billion dollars in forty-five minutes because its trading code had never sparred with a live market that could retaliate at millisecond speed. One example turns rolling on gym mats into global dominance; the other converts an "all tests pass" memo into existential ruin.

## Positive Example — Brazilian Jiu-Jitsu's Ascendance in Early Mixed Martial Arts

This is perhaps one of my favorite stories that emerged in my study of adversarial dynamics. The tale begins not in Las Vegas but in Rio de Janeiro's humid garages during the 1920s. Carlos Gracie, a young judo enthusiast, tweaked Kodokan throws for street fights and taught his brothers: notably the wiry Helio Gracie, who lacked brute strength but made up for it with leverage tricks on the ground. Their laboratory was the *vale tudo* circuit—no-glove, no-time-limit challenge matches where one could head-butt, soccer-kick, or choke at will. Technique forged there passed only if it survived genuine chaos.

Fast-forward to November 12, 1993. Pay-per-view audiences tuned in to the first Ultimate Fighting Championship expecting blood-sport spectacle. UFC was a dream-come-true for every kid of the 70s and 80s who engaged in schoolyard arguments about which martial art was "best." UFC was—by design—meant to be interdisciplinary, pitting the top practitioners in each fighting art against each other to see if any particular discipline emerged victorious. By all appearances, Royce Gracie seemed most unimpressive. He entered at 176 pounds wearing a baggy gi. What viewers saw shocked the martial-arts establishment: Royce clinched, tripped, and choked three opponents in five minutes of combined cage time even though he threw no strikes harder than a slap. By UFC 4 he had finished Dan Severn—a 260-pound NCAA champion wrestler—using a triangle choke unheard of in most gyms.

Why did BJJ succeed so emphatically? Because its practitioners spent their lives in adaptive pressure tests. Classes typically split time between technique and live *rolling*—unscripted grappling rounds where partners resist full tilt. Mistakes lead to joint locks or strangulation, but a quick tap resets the drill. Failure becomes data, cycled back into the next roll. A black belt—notoriously hard to achieve in BJJ—amasses ten thousand such micro-failures, each refining timing and leverage.

Traditional martial artists of the era often drilled forms or one-step sparring—punch, block, counter—against compliant partners. High kicks looked spectacular in mirrors but crumbled when an opponent jammed distance, clinched hips, and dragged the fight to canvas. BJJ's ground paradigm flipped the game board: strikers found themselves mounted, punches stifled, chokes tightening.

The ripple effect was immediate. Through the mid-to-late 1990s, Kenpo and other stand-up schools added grappling electives; Muay Thai fighters increasingly drilled sprawl-and-brawl defenses; and by the 2000s most serious MMA gyms scheduled as much mat time as pad work. Traditional disciplines also adapted: judo rulesets put greater emphasis on newaza in periods, and some wrestling events experimented with submission divisions.

Royce's later record was mixed once rivals cross-trained, but that is evolution's hallmark: the organism that first embraced adaptive testing forced the environment to change. BJJ's triumph cemented a truth: a style stress-tested daily against non-compliant partners scales under lights better than a style honed only in cooperative drills. Promoters and coaches widely credit the early Gracie dominance for modern MMA's training renaissance—live rounds, cage-wall wrestling, interdisciplinary sparring.

Beyond cages, the lesson migrated. SWAT teams replaced static shooting lanes with force-on-force Simunition® scenarios. Software startups replaced waterfall releases with canary deploys that pit new code against real users in micro-slices. Across contexts, BJJ's garage-laboratory ethos—fail small, learn fast, iterate under pressure—rewired preparation culture. And we're all better for it.

### Negative Example — Knight Capital's 45-Minute Meltdown

Knight Capital was, by mid-2012, the largest U.S. equities market maker, responsible for routing nearly 17% of NYSE trades each day. Speed and trust were its currencies. On August 1, at 9:30 a.m. Eastern, both evaporated. Quickly.

Developers had spent months crafting enhancements to Knight's Smart Market Access Routing System, nicknamed SMARS. The new sub-module, *Power Peg*, aimed to detect micro-price discrepancies and inject liquidity before rivals could pounce. On July 31, technicians copied executables to eight production servers, but an oversight left one server running obsolete code; when a new software feature *repurposed an old command flag*, that single server misinterpreted the signal and activated the dormant, high-volume trading instructions. When the opening bell clanged on August 1, that lone server activated the flag and began firing a torrent of orders, buying high and selling low.

In less than a minute, abnormal volumes rippled across over a hundred ticker symbols. Adaptive resistance kicked in. High-frequency trading firms smelled mispricing, shorted the anomalies, and filled their books with risk-free profits. NYSE's order-imbalance guards, designed to reopen a ticker after extreme moves, tripped repeatedly but were too slow to staunch the flow.

Knight's monitoring dashboards lit up, but alerts drowned staff in noise: normal busy-morning spikes obscured the pattern. At 9:45 a.m.

a senior engineer yelled across the command center, "Kill all SMARS!" Trades kept firing; the rogue server ignored shutdown commands, still reading its legacy flag.

By 10:15 a.m. Knight's net positions showed approximately a $440 million loss. In forty-five minutes the firm had accumulated a loss equal to four times its second-quarter earnings. Risk managers contacted clearing firms seeking unwind relief, but counterparties declined—profits were locked and legal to keep. News desks ran "Knightmare on Wall Street." Knight's share price cratered 75% by lunch.

Postincident analysis revealed astonishing gaps in adversarial validation. *Power Peg* had passed user-acceptance tests against two weeks of sanitized historical data. No sandbox unleashed the code against live-engine stubs capable of spoofing contradictory fills. There were no chaos drills yanking cables mid-session, no kill-switch rehearsals. One internal document described testing as "exact replay"—the antithesis of adaptive resistance.

Regulators fined Knight $12 million, modest compared with losses, but the reputational damage was mortal. CEO Thomas Joyce scrambled for capital, eventually accepting a $400 million rescue from a Goldman-led consortium at punitive terms. In July 2013 Knight merged into rival GETCO; the iconic orange trading jackets disappeared.

Knight's fall is a parable of sterile validation. A single server, over-looked during deployment, carried code from an era when the firm still used eight-second quote intervals. Confronted with a live market that punishes error at light-speed, the code hemorrhaged cash faster than humans could react. Unlike BJJ gyms where students tap early and learn, Knight's sandbox never choked the code into revelation. The first real resistance was terminal.

## Principles in Action — Reflection Questions

1. Does our testing environment empower independent aggressors to invent brand-new attack paths, or merely replay last month's failures?
2. When our red team succeeds, do we reward the find and ship a fix inside a sprint, or bury the embarrassment in a postmortem PDF?
3. Which dormant settings, flags, or rare-path code branches could resurrect in production, and have they ever faced a chaos drill?
4. Do frontline operators train under the same time pressure, ambiguity, and sensory overload they will face when the system falters?
5. Where is the ethical red line for our adversarial testing, and who owns the kill-switch if realism veers toward recklessness?

## Sources

- Amazon Web Services. 2017. "GameDay: Simulating Real-World Outages to Build Resiliency." *AWS Architecture Blog,* November 14, 2017. http://aws.amazon.com/blogs/architecture/gameday-simulating-real-world-outages-to-build-resiliency/
- Gentry, Justin. 2007. *The Ultimate Fighting Championship: A History of the UFC's First 100 Events.* Toronto: ECW Press.
- Gracie, Rorion. 1994. "The Birth of Brazilian Jiu-Jitsu." Interview. *Black Belt,* July 1994.
- Grant, Justin. 2012. "Knightmare on Wall Street: A Lesson in Risk Management." *Financial Times,* August 3, 2012.
- Liddell, Chuck, and Chad Dundas. 2008. *Iceman: My Fighting Life.* New York: Dutton.

- Netflix. n.d. "The Chaos Monkey Guide to Testing." Netflix. http://netflix.github.io/chaosmonkey/
- Securities and Exchange Commission. 2013. "In the Matter of Knight Capital Americas LLC, Respondent." Administrative Proceeding No. 3-15570, October 16, 2013.
- Seneca the Younger. 1917. *Moral Letters to Lucilius,* Letter 13. Translated by R. M. Gummere. Cambridge, MA: Harvard University Press.
- SpaceX. 2021. "Starship Testing Overview." SpaceX, March 2021. http://www.spacex.com/vehicles/starship/
- The Wall Street Journal. 2012. "A Tale of Human Error at Knight." *The Wall Street Journal,* August 2, 2012.

# PRINCIPLE 9
## ACCOUNT FOR INSIDER RISK

*"The enemy is within the gates."* — Marcus Tullius Cicero

The higher the fence, the more dangerous the gatekeeper. External attackers struggle against visible obstacles; insiders step over them with a badge swipe and a smile. That asymmetry explains why IBM Security's 2023 Cost of a Data Breach report found the average breach cost in the millions of dollars—and insider-driven incidents often take months to identify and contain. Those numbers symbolize more than money. Across domains, they may mark canceled product launches, executed intelligence agents, and patients who wonder why their medical records are on the dark web.

To manage insider risk, we have to see it clearly. We'll use four practical pathways—the "Four M's"—and map each to a layer of defense: people, process, platform, and culture.

**Malicious insiders** chase money, ideology, or revenge. They show up in sudden wealth spikes, hostile social-media posts, or secret job searches. These are best countered with continuous monitoring, strict access controls, and fast-acting investigative processes that can pivot to HR or law enforcement.

**Manipulated insiders** buckle under blackmail or romance scams. Their footprints appear in whispers in chat logs, unexplained schedule swaps, or odd travel bookings. Process safeguards—like multi-person approval for high-impact actions—and platform monitoring that correlates behavior with external triggers help contain the damage.

**Mistaken insiders** simply click the wrong link or mistype a command. Not to disagree with Cicero here, but sometimes a nation can't survive its fools. This is where people-focused measures matter most: targeted training, prompt rollback procedures, and blame-free hotlines that encourage rapid reporting before errors snowball.

**Marginalized insiders** feel small, unheard, or mocked. Their warning signs appear in engagement surveys and hallway grumbles long before they flip a server switch. Cultural countermeasures—psychological safety, visible mission alignment, and leaders who act on grievances—are the best insulation against this category.

No single tool blocks every path, but when each "M" is paired with the right mix of people, process, platform, and culture controls, the combined layers can drastically reduce the odds that an insider can—or will—do lasting damage.

∼

PEOPLE CONTROLS START WITH CLEAR MISSION. IF AN EMPLOYEE CAN SAY in one breath why their work matters, betrayal feels like self-harm. NASA's "safety over schedule" mantra allowed engineers to halt dangerous shuttle countdowns; hospitals that open staff huddles with patient stories report fewer insider pill thefts. Psychological-safety research backs the point: teams that feel free to raise concerns suffer fewer hidden errors.

**Process controls** add healthy friction. Mandatory vacations force privileged users offline so anomalies surface. Quarterly tabletop drills begin with a devastating assumption: *the attacker already has credentials.* Red teamers then test whether alerts fire, whether SOC staff are authorized to shut down the session, and whether legal, HR, and engineering know who speaks first to the board.

**Platform controls** cover least-privilege access, just-in-time elevation, and outbound data-loss prevention. Modern behavioral-analytics tools do more than count downloads; they model normal patterns. When a procurement clerk who works nine-to-five suddenly uploads gigabytes at 2 a.m., the system pings analysts in minutes.

But controls fail if culture mutes the message. Managers who punish every false alarm teach staff to sit on suspicions until after the breach is public. The best programs celebrate "near misses." When a new hire flags an odd query and it turns out benign, leadership still thanks them at the next stand-up. That applause rewires peer pressure in favor of vigilance.

~

INSIDER-RISK MATURITY OFTEN FOLLOWS FOUR PLATEAUS:

1. *Reactive.* Incident response only; breaches discovered after the fact.
2. *Basic monitoring.* Admins run daily log scans; alerts email to an overloaded inbox.
3. *Integrated.* Logs, HR data, and badge events feed a common dashboard; SOC has lockout authority.
4. *Predictive.* Machine learning ranks risk, red-team drills validate, and culture rewards early reporting.

Most organizations hover between stages two and three. Moving up demands both budget and courage. Budget buys software; courage lets security freeze the vice president's account at 3 a.m. without an angry phone call halting progress.

The 2013 Edward Snowden disclosures revealed a similar gap. While the NSA deployed some of the world's most advanced digital monitoring, Snowden himself later noted that the physical security was often lax enough that he could carry microSD cards out of the facility hidden in his pocket. That gap between digital policy and physical reality is what modern insider threat programs aim to close. They can use techniques like "canary spreadsheets"—files with invisible beacons—to catch an employee opening sensitive data; an alert can arrive in minutes, allowing HR to conduct an exit interview before lunch.

## Positive Example — Tesla Detects and Stops Sabotage at the Gigafactory

In early 2018 Tesla's future hinged on a single number: 5,000. That was the weekly quota of Model 3 battery packs the company promised to reach by the end of June. If the target was met, analysts would finally see the electric-car maker as a mass-producer; if it was missed, critics could claim the business would never scale. Inside Gigafactory 1 just east of Reno, engineers raced twenty-four hours a day to fine-tune robots, trim scrap, and keep lithium-ion cells flowing into steel casings. Pressure of that sort can distort judgment. A tired operator may bypass a safety check to keep the line moving; a disgruntled employee may decide to embarrass management by leaking data; an opportunist may try to profit by handing proprietary designs to short-selling traders. Tesla's leadership knew those temptations well, and in late 2017 it had doubled down on insider defenses: every workstation on the manufacturing network reported telemetry to a central dashboard; data-loss-prevention software scanned outbound traffic for encrypted archives; badge readers posted time-stamped swipes to a live map; and the Security Operations Center had standing authority to disable any user account the moment something looked wrong.

*Details that follow reflect Tesla's contemporaneous logs, Elon Musk's June 2018 company-wide email, and the company's federal court filings; Tripp publicly disputed Tesla's characterizations.*

Those layers were put to the test on the quiet shift between Saturday night, June 16, and Sunday morning, June 17. Overnight, the

DLP console flashed red: large archives of production data were leaving the network from a station that normally handled only scrap-yard totals. Seconds later a different alert showed that the same work-station had executed a script against the Manufacturing Operating System, altering database queries so that they appeared to come from other users. The SOC analyst on duty followed the escalation play-book: first freeze the login, then trace the badge history. Logs revealed that the owner of the account, process technician *Martin Tripp*, had just clocked out. With credentials locked and no one physically at the desk, the risk of further exfiltration fell to zero, buying investigators the daylight they needed.

By mid-morning forensic staff had cloned Tripp's hard drive and opened the ZIP bundles. Inside were photos of battery-module assembly lines, scrap-rate spreadsheets, and portions of proprietary design drawings. A Python script was running on the station; Tesla later told the court the code siphoned production data and that public claims drawn from that data overstated scrap and implied safety issues the company disputed. Had the code run undetected for a few days, Tesla's yield dashboard could have convinced managers that pack quality was falling apart, possibly leading them to halt production during the make-or-break month. At this point the insider-threat plan shifted from containment to notification. Legal, human resources, and the battery-engineering lead convened in a small conference room and walked through a short checklist: *Evidence chain preserved? Yes. Operational risk mitigated? Yes. External disclosure needed?* Two hours later, at 2 p.m. Pacific, employees received an email from CEO Elon Musk. In characteristic plain language he wrote that *"a trusted employee conducted quite extensive and damaging sabotage,"* but production systems were secure and authori-ties had been contacted.

Three days later, Tesla filed a civil complaint in U.S. District Court, accusing Tripp of hacking MOS, exporting gigabytes of trade secrets, and making false statements to the media. The filing cited network logs and sworn forensic reports; because the SOC had reacted in minutes, all digital evidence lined up cleanly. Authorities were notified. In September 2020 the court largely ruled in Tesla's favor, and in December 2020 a filing recorded a $400,000 payment by Tripp to resolve the case. Operational fallout at the factory was limited to a single overnight pause while engineers verified robot settings. The stock price barely twitched, and by the last week of June the line hit 5,031 packs—*goal met.*

What makes the incident a teaching case is not luck but design. First, detection did not hinge on a single tool. Telemetry, DLP scanning, and badge analytics overlapped so that failure of one sensor would still leave others to ring alarms. Second, authority was clear: the SOC did not need vice-presidential approval to freeze credentials; speed trumped hierarchy. Third, rehearsed choreography joined security, legal, and engineering in hours, not days. Quarterly tabletop drills had walked those players through the same script in low-stress settings, so each knew where to sit and what to sign when a real alert arrived. Finally, Tesla chose transparency over secrecy. Musk's company-wide email and subsequent public complaint framed the discovery as proof the system worked, defusing rumor before it could erode morale.

The contrast with historic failures such as Aldrich Ames is sharp. At Tesla, unusual network traffic surfaced within minutes; at the CIA, unexplained cash and strange file access sat unchallenged for nearly a decade. Tesla's safeguards show that insider risk is manageable when technology, policy, and culture aim at the same goal: *detect early, act fast, and keep the mission larger than any one person's frustration.*

### Negative Example — Aldrich Ames and the CIA's Darkest Hour

Aldrich Ames looked ordinary enough—slouched posture, rumpled suit, brown bag lunches spiked with bourbon. Yet from 1985 to 1994 he gutted the CIA's Soviet network. Ames's slide into betrayal began with money pressure and pride. After a costly divorce and a new marriage to María del Rosario Cano, he craved the image of success his stalled career could not provide. He felt resentful towards the CIA for his situation. On April 16, 1985, he entered the Soviet Embassy and, in the space of a few handshakes, sold more secrets than the Rosenbergs and John Walker combined.

Within three months Moscow had arrested or executed six CIA sources. Ames continued feeding names through dead drops labeled by code words that only he and the KGB handlers understood. He passed two polygraph exams thanks to beta-blockers that flattened his heart-rate spike. Meanwhile, the Jaguar in the parking lot and cash paid for a Georgetown home went largely unchallenged. HR flagged the wealth anomaly, but finance believed the story of María's "family money," and security considered the matter outside their lane. No single office saw enough of the puzzle to shout stop.

What finally cracked the case was a cross-functional mole-hunt team given authority to break silos. Analysts connected Ames's data pulls, foreign-currency bank deposits, and Moscow's sudden success against U.S. assets. Trash pulls revealed $300 bottles of wine; a surveillance team watched Ames drop a chalk mark on a D.C. mailbox—the prearranged signal for a KGB meet. On February 21, 1994, FBI agents boxed his car at a traffic light. The final damage assessment was catastrophic: the betrayal led to the execution of at least ten of the CIA's highest-level Soviet sources and the compromise of countless operations, leaving the agency's intelligence network against its primary adversary in ruins.

In the aftermath, Congress blamed three gaps: mission drift that let personal resentment fester (Remember Principle 2?), a culture that prized autonomy over accountability, and technical tools—polygraph, financial disclosure—that either failed or no one linked together. If the anger of one employee is a match, those gaps were the gasoline.

### Principles in Action — Reflection Questions

1. Have we charted malicious, manipulated, mistaken, and

marginalized pathways for each sensitive role and tied a control to every path?

2. Do our monitoring systems route critical alerts to people empowered to freeze access within minutes, not hours?

3. Can any employee raise a suspicion safely, and do leaders publicly thank them even when the alert is a false alarm?

4. How often do red-team drills start with the assumption that an attacker already has valid credentials and knows the jargon?

5. When hitting a deadline conflicts with living the mission, which one wins at two o'clock in the morning?

## Sources

- Carnegie Mellon University, CERT Division. 2023. *Insider Threat Center Glossary.* Pittsburgh, PA: Carnegie Mellon University.
- Central Intelligence Agency. 1995. *Ames Damage Assessment Report.* Declassified summary.
- Cicero, Marcus Tullius. 1976. *In Catilinam I–IV.* Translated by C. MacDonald. Cambridge, MA: Harvard University Press.
- Harding, Luke. 2014. *The Snowden Files: The Inside Story of the World's Most Wanted Man.* New York: Vintage Books.
- IBM Security. 2023. *Cost of a Data Breach Report 2023.* Armonk, NY: IBM.
- Kolodny, Lora. 2018. "Tesla Sues Ex-Gigafactory Technician, Alleging Hacking and Sabotage." *CNBC,* June 21, 2018.
- Musk, Elon. 2018. Internal email to employees, "Some Concerning News," June 18, 2018.
- Prunckun, Hank. 2019. *Counterintelligence Theory and Practice.* Lanham, MD: Rowman & Littlefield.
- SANS Institute. 2023. *Common Insider Threat Patterns.*

- Tesla, Inc. v. Tripp, No. 3:18-cv-00293-MMD-WGC (D. Nev. December 10, 2020).
- U.S. Senate Select Committee on Intelligence. 1994. *Statement on the Aldrich Ames Espionage Case.* March 1994.
- Weiner, Tim, David Johnston, and Neil A. Lewis. 1995. *Betrayal: The Story of Aldrich Ames, an American Spy.* New York: Random House.

# PRINCIPLE 10

## ENGAGE STAKEHOLDERS WITH INTENTIONALITY

*"Four hostile newspapers are more to be feared than a thousand bayonets."*
— Napoléon Bonaparte

You don't need to be outgunned to lose. You can have strategy, capability, initiative, and cohesion. But if the people who control access, legitimacy, or permission aren't with you—or weren't even told it was happening—you're dead on arrival.

Every engagement contains three types of people. There are those who fight for you: your team, your unit, your staff, your institution. There are those who fight with you: your allies and formal partners, whether on the battlefield or in the boardroom. And then there are those who don't fight at all—but whose decisions shape the entire terrain of conflict.

Those are your stakeholders.

They don't wear your uniform. They don't carry your flag. They don't swing at your adversary. But they hold power in ways that matter. Oversight stakeholders—like boards, funders, political appointees, and senior executives—control the resources and institutional runway that make engagement possible. Gatekeeping stakeholders—like regulators, community leaders, certifiers, international observers, and even algorithms—can grant or withhold legitimacy. They decide whether you're allowed to move forward, and on what terms.

Stakeholders are not the ones who show up when the fight begins. They are the ones who decide whether the fight is allowed to begin at all.

And yet they are too often ignored. Leaders plan in secret. Advisors brief internally. Timelines are set in rooms without a single stakeholder present. Then, when engagement is imminent, someone sends an email. Or schedules a press release. Or makes a call a week too late. And the stakeholder who was never involved—never asked what they wanted, never heard what they feared—says no. Or hesitates. Or forwards the request to legal. And suddenly the op stalls.

Stakeholder engagement isn't a box to check. It's a system of influence. And if you don't design it with *intent*, it doesn't exist.

Mapping is the first discipline. Every stakeholder must be named —not by title, but by *person*. "The press" is not a stakeholder. "The licensing board" is not a stakeholder. People are. What are their names? What power do they hold? How interested are they in the outcome? If you can't answer those questions, you aren't ready to act.

Once mapped, you define their motivations:

- They *want* ...
- They *fear* ...

You write these two sentences for every stakeholder. Then you go validate them. You ask questions. You shut up and listen. You test assumptions. Because if you're wrong, it's better to find out now than after the gates are closed.

Then you engage. Intentionally.

Before engagement, stakeholder communication is about surfacing concerns, building trust, and designing mutual benefit. It's quiet. It's relational. It's often informal. But it sets the stage.

During engagement, communication must shift from persuasion

to transparency. You don't pitch. You inform. Stakeholders need liaisons, dashboards, visibility, and responsiveness. And if your internal cadence doesn't match their risk tolerance, your engagement rhythm is wrong.

After engagement, the loop must close. Stakeholders want to know: Did it work? Did it deliver? Did their support matter? If you don't answer those questions, you may not get another yes.

Rewards are the hidden lever. Stakeholders must see a clear upside for their involvement—whether it's reputational, political, financial, operational, or personal. If they put their name behind you and get nothing in return, you lose them. And if they feel used, forgotten, or ghosted, they will never come back.

And when the facts change—as they always do—you must recalibrate. Stakeholder wants and fears can shift overnight. If no one on your team owns the stakeholder narrative—if no one is authorized to push updates across the grid within 24 hours—you're already behind. And you won't know it until it's too late.

## Positive Example — NASA's Artemis Program

By the late 2010s, NASA's human spaceflight program was adrift. The Space Shuttle had been retired for years, and subsequent projects had been canceled or delayed, leaving a vacuum in American space leadership. The Artemis Program was designed to fix this. Its mission: to establish the first long-term human presence on the Moon, creating a sustainable deep-space architecture that could one day take astronauts to Mars. But the technical challenge was accompanied by an equally challenging political one: after years of shifting goals and budget fights, NASA had to prove it could manage a massive coalition of powerful stakeholders to make the vision a reality.

From the outset, Artemis treated stakeholder management as a design requirement, not an afterthought. NASA addressed what Congressional appropriators *wanted*—jobs, contracts, and STEM visibility for their districts—while soothing what they *feared*: being seen as funding a rudderless program with no tangible local benefit. Every funding request was bundled with tangible, localized proof of value.

Commercial and international partners *wanted* reliable schedules and a clear return on their massive R&D investments; they *feared* getting stuck in a political quagmire where requirements changed with every election cycle. To counter this, NASA didn't just hand them timelines—it made them contributors. NASA gave the FAA and European Space Agency access to its internal scheduling and readiness dashboards. Delays were not hidden; they were discussed in shared forums with stakeholder buy-in. When a component slipped, the reason wasn't buried—it was communicated with data, diagrams, and new dates.

Even the general public was treated as a stakeholder. NASA streamed SLS rocket tests live. TikTok videos from young engineers showcased behind-the-scenes progress. Teachers received curriculum packets and printable models for classrooms. The message was clear: Artemis is *yours*, too. And when the first Artemis launch slipped by nearly a year, support didn't crumble. Budgets weren't pulled. Journalists didn't frame it as failure. Partners didn't walk.

Because expectations had been set. Because legitimacy had been earned. Because the stakeholders had been engaged.

To be clear, the ultimate success of Artemis—a sustainable human presence on the Moon—is still years away. But the success of its stakeholder engagement is already evident. By securing the long-term political, commercial, and public support necessary to survive inevitable delays and technical hurdles, NASA achieved its most critical near-term objective: building a coalition resilient enough to see the mission through.

Artemis didn't succeed because it avoided friction. It succeeded because it anticipated friction and brought everyone to the table long before it arrived.

**Negative Example — The European Super League: A Rebellion Crushed by Its Stakeholders**

In April 2021, twelve of the wealthiest and most powerful football clubs in Europe executed a secret plan to form a breakaway "Super League." This was a direct, adversarial attack meant to seize control of European football from its longtime governing body, UEFA. The clubs, armed with billions in financing from JPMorgan Chase, prepared for a legal and political war with UEFA. But they completely failed to engage the stakeholders whose loyalty they took for granted, leading to the project's spectacular collapse in less than 48 hours.

The plan's most catastrophic failure was its handling of the most important gatekeeping stakeholder: the *fans*.

- The club owners, operating in secret, fundamentally misunderstood their own supporters. They believed the fans' *want* was simply to watch their team play against other big teams, regardless of context.
- They failed to recognize the fans' deeper *fear*: the destruction of the 100-year-old tradition of meritocracy, promotion, and relegation that forms the soul of European football. The "Super League" was a closed shop, an American-style franchise model where the founding members could never be kicked out, no matter how poorly they performed.
- The backlash was not just negative; it was immediate, unified, and volcanic. Fans of rival clubs protested together, unfurling banners at stadiums that read "Created by the Poor, Stolen by the Rich." They successfully framed the conflict not as a business dispute, but as a hostile, illegitimate theft of their cultural heritage.

The organizers also misjudged their internal stakeholders: the *players* and *managers*.

- They failed to map their motivations, assuming employees would fall in line. The players and managers *wanted* to compete for historic trophies like the World Cup and Champions League; they *feared* being banned from those very competitions and being branded as mercenaries by the fans whose support they depend on.
- As a result, influential figures like Liverpool's Jürgen Klopp and Manchester City's Kevin De Bruyne publicly condemned the plan. By failing to secure the consent of their own uniformed "soldiers," the club owners lost all legitimacy on the field.

Finally, they ignored the power of *government gatekeepers*.

- The club owners wanted a free market where they could operate without interference; they feared political intervention. But the UK Government *wanted* to protect a beloved cultural institution and *feared* the voter backlash of appearing to side with billionaire owners against the fans.
- This misalignment was critical. The British Prime Minister immediately threatened to drop a "legislative bomb" to stop the breakaway. The clubs were prepared for a fight with UEFA, but not with their own national governments.

The result was one of the most stunning failures in modern sports business history. The project imploded not from an attack by its adversary, UEFA, but from the overwhelming power of its stakeholders. One by one, the clubs withdrew, issuing groveling apologies to their fans.

The Super League founders thought they were in a boardroom battle for money and power. They failed to realize their stakeholders believed the conflict was about identity, community, and history. It is a perfect, modern example of how an adversarial campaign, no matter how well-funded, is dead on arrival if you fail to understand and engage the people who grant you the legitimacy to act in the first place.

## Principles in Action — Reflection Questions

1. Have we explicitly mapped every stakeholder with power to grant or block permission, access, legitimacy, or resources?
2. Do we understand their wants and fears—and have we validated those answers directly?
3. Are our communication rhythms aligned to stakeholder risk, not just our own operational tempo?
4. Do stakeholders see visible, timely rewards tied to their involvement?
5. When facts shift, who owns the stakeholder narrative— and how fast can they update it?

## Sources

- Bonaparte, Napoléon. 1966. Quoted in J. L. Heinl Jr., ed. *Dictionary of Military and Naval Quotations.* Annapolis, MD: United States Naval Institute.
- Conn, David. 2021. "'A Grotesque Betrayal': How The Guardian Reported the Super League Story." *The Guardian,* April 23, 2021. http://www.theguardian.com/ football/2021/apr/23/a-grotesque-betrayal-how-the-guardian-reported-the-super-league-story
- Government Accountability Office. 2024. *NASA: Assessments of Major Projects.* Washington, DC: GAO.
- NASA. 2023. *Artemis Program Plan.* Washington, DC: NASA. http://www.nasa.gov/wp-content/uploads/2023/09/artemis-program-plan-2023.pdf

- Panja, Tariq, and Rory Smith. 2021. "How the Super League Fell Apart." *The New York Times,* April 22, 2021. http://www.nytimes.com/2021/04/22/sports/soccer/super-league-soccer-florentino-perez.html
- Rumsby, Ben. 2021. "'A Legislative Bomb': How Boris Johnson Helped Blow Up the European Super League." *The Telegraph,* April 21, 2021. http://www.telegraph.co.uk/football/2021/04/21/legislative-bomb-boris-johnson-helped-blow-european-super-league/
- Wilson, Jonathan. 2021. "The Core Problem of the Super League: A Lack of Soul." *Sports Illustrated,* April 19, 2021. http://www.si.com/soccer/2021/04/19/european-super-league-soul-tradition-history-greed-liverpool
- Ziegler, Martyn, and Matt Lawton. 2021. "European Super League: All Six Premier League Clubs Begin Withdrawal Process." *The Times,* April 20, 2021. http://www.thetimes.co.uk/article/european-super-league-all-six-premier-league-clubs-begin-withdrawal-process-f6botoqf9

# PRINCIPLE 11

## SHAPE THE ENVIRONMENT TO YOUR ADVANTAGE

*"The true aim is not so much to seek battle as to seek a strategic situation so advantageous that if it does not of itself produce the decision, its continuation by a battle is sure to do so." — B. H. Liddell Hart*

Most people picture conflict as a head-to-head collision: armies rushing across fields, corporations battling for market share, activists sparring in comment threads. Yet long before the opening clash, the ground itself can be tilted. *Shape the Environment to Your Advantage* insists that deliberate terrain work—*physical, procedural, informational,* and *psychological*—creates immense leverage.

Begin with the obvious: *physical* space. A single hill can halve the manpower needed to guard a frontier; a server room tucked behind two locked doors and a ninety-degree blind turn buys incident-response teams the minutes they need after an alarm. Physical shaping is old, but its logic is evergreen: make your movements easy and your adversary's exhausting.

Below the surface lies *procedural* space—the mesh of bylaws, licensing terms, zoning rules, and protocol headers that either widen or narrow lanes of action. One sentence in a procurement statute can exclude firms that lack a particular certification; a three-line patent claim can force rival products to redesign around your geometry at ruinous cost. Procedural shaping is quiet and durable: once a rule calcifies, it polices itself.

Flooding every corner is *informational* space. What people notice, search, and share bends the arc of conflict as surely as trenches. A daily press briefing timed for dawn anchors every newsroom's agenda; a lightweight file format offered free of royalties can push competing standards off the screen. When information flows have been routed through your gatehouses, each data packet reinforces the tilt you engineered.

Finally comes *psychological* space—the climate of expectations and emotions in which actors interpret every signal. A team convinced that the plan is inevitable approaches tasks with speed and poise; a rival who senses the deck is stacked tires faster, doubts longer, and risks less. Conditioning this climate can be as subtle as framing a policy change as "common sense" rather than "experimental," or as vivid as staging an awe-inducing product reveal in a venue designed to dwarf skepticism.

These four layers rarely operate in isolation. Apple doesn't just launch a product. Months before the reveal:

- They file strategic patents and compliance documents (procedural)
- They redesign retail spaces and update UI mockups across devices (physical)
- They benefit from (and are widely suspected of cultivating) a culture of controlled leaks that builds anticipation through supply-chain blogs (informational)
- They reinforce a sense of inevitability that "something revolutionary is coming" (psychological)

By the time the keynote airs, the audience is already primed to believe. Journalists have narratives half-written, competitors are playing defense, and consumers feel like not buying is missing out.

That is shaping: one coordinated effort, cascading across layers, to preconfigure the environment in your favor.

When planning a shape-the-environment campaign, ask three questions:

1. What future move should feel effortless for me?
2. What future move should feel cumbersome for my opponent?
3. How can a single intervention cascade across multiple layers rather than lodging in just one?

Shaping is not pure offense; it demands stewardship. Ignored moats breed mosquitoes that weaken the garrison, and rule sets carved for yesterday's threats can trap their authors tomorrow. Sustainable advantage requires gardeners, not just architects: people who revisit the earthworks, patch narrative leaks, update standards, and refresh the sense of inevitability that keeps morale high.

The payoff is asymmetric. Once the field slopes your way, every unit of energy, budget, or courage you expend travels farther, while every unit the adversary spends climbs uphill. Battles may still be difficult, but they unfold on your timetable, at costs you predicted, against opponents already forced into awkward shapes by the very ground beneath them.

### Positive Example — The NAACP's Terrain-Shaping Campaign Before *Brown v. Board of Education*

When Charles Hamilton Houston began building a legal strategy to dismantle segregation in 1929, he understood that a direct assault on *Plessy v. Ferguson* would almost certainly fail. The doctrine of "separate but equal" had calcified into law and culture for over three decades. The terrain was not favorable. So Houston and his successor, Thurgood Marshall, chose not to rush the hill—but to *reshape* it.

Over the next 25 years, the NAACP executed a deliberate campaign to tilt the environment across three key layers: *procedural, informational, and psychological.* The fourth layer, *physical*, played little to no role—which itself underscores an important lesson: shaping does not always require full-spectrum dominance, only the right leverage points for the domain.

*Procedural shaping* was the first priority. In case after case, the NAACP narrowed the legal space that segregationists could operate in. *Gaines v. Canada* (1938) forced Missouri to provide an in-state legal education for Black students, eliminating the common procedural dodge of out-of-state tuition reimbursement. *Sweatt v. Painter* (1950) required courts to consider qualitative aspects of education—like institutional reputation and alumni networks—when assessing equality. These rulings didn't end segregation, but they methodically constricted the range of legally viable excuses for maintaining it. With each precedent, the NAACP rewrote the rules of engagement.

*Informational shaping* took place in the broader public arena. The NAACP worked strategically to frame segregation not just as a legal oddity, but as a moral contradiction at the heart of American democracy. Through curated media coverage, academic reports, and coordinated amicus briefs, they flooded the discourse with data and stories that reframed separate schooling as both inferior and un-American. By amplifying these narratives across newspapers, radio, and academic channels, they influenced what the public—and eventually the courts—focused on. This wasn't just advocacy; it was a deliberate campaign to saturate the informational environment with a new default assumption: that segregation was indefensible when measured against the nation's professed ideals.

*Psychological shaping* worked more subtly, altering the emotional and anticipatory climate in which decisions were made. The NAACP timed their legal actions to coincide with broader societal shifts: postwar liberalization, emerging Cold War contradictions, and a growing global spotlight on American race relations. The Clark doll studies, often cited for their evidentiary value, also played a role in psychological terrain-shaping—not because they proved harm in a clinical sense, but because they made that harm emotionally legible. Over time, segregation was increasingly framed as legally untenable and morally unsustainable, especially under the Cold War spotlight. That atmosphere of inevitability—the quiet erosion of moral and institutional confidence in the segregationist position—shaped how judges, journalists, and the public interpreted every new development.

*Physical shaping*, by contrast, was largely absent. The NAACP didn't control buildings, school locations, or geographic access to facilities. This wasn't a campaign fought with infrastructure. But that absence is revealing. In legal and cultural struggles, informational and psychological leverage can often outweigh terrain made of brick and concrete.

By the time *Brown v. Board of Education* reached the Supreme Court in 1952, the ground beneath *Plessy* had already been shifted. The lanes of argument had narrowed. The dominant narrative had changed. The emotional climate had tilted. When the Court ruled in 1954 that separate schools were inherently unequal, it felt like a breakthrough to many. But for Houston and Marshall, it was simply the final step on a slope they had been grading for a generation. The ruling didn't need to climb—it merely needed to follow gravity.

## Negative Example — Record Labels and the Ungraded Terrain of Digital Music

At the turn of 1999, Universal, Sony, Warner, EMI, and BMG were earning billions from compact discs. The terrain felt defensible. Copyright law favored them. Physical media and distribution channels were expensive to produce and tightly controlled. The threat of digital piracy, enabled by technologies like MP3 compression, was no secret, but the labels saw it as a fringe hobby—not a force requiring a reshaping of the battlefield.

Rather than *shape* the emerging digital environment, they tried to preserve the old one. They filed lawsuits. They redesigned jewel cases. They leaned on traditional marketing muscle. But they built no licensing infrastructure, no cross-label storefront, no digital-native payment system. The *procedural layer*—the rules and frameworks that would govern digital music—was left undeveloped, unguarded, and unclaimed.

On June 1, 1999, a teenager in a dorm room released Napster. The software redefined the *physical layer*. By shifting distribution to a peer-to-peer architecture, it bypassed U.S. borders and legal enforcement entirely. Files flowed from user to user across international terrain, eroding any advantage the labels once held from controlling factories, trucks, or shelves. The map changed. The labels stayed put.

Napster also reconfigured the *informational layer*. Its interface turned music into a searchable, shareable, frictionless experience. Users browsed by artist, song, or playlist—curated by peers rather than corporations. Yet while Napster won attention, the labels still held the upper hand in narrative legitimacy: they had the law, the history, and the moral high ground. But they never used it. No campaign explained the human cost of piracy. No effort reframed free downloads as theft from working musicians. The informational battlefield lay open, and the labels never planted a flag.

Their greatest failure, though, was in the *psychological layer*. For a brief window, they still owned the emotional frame—trusted brands, household names, and cultural capital. Had they offered a clear, fair digital path forward, consumers might have followed. Instead, they appeared defensive and out of touch. Lawsuits against college students created backlash, not deterrence. Napster offered novelty and empowerment. The labels appeared punitive and defensive. Expectations hardened. Legitimacy withered. The sense that "music should be free now" filled the vacuum they left unaddressed.

Apple recognized what the labels missed. On April 28, 2003, it launched the iTunes Music Store with 99-cent tracks. The files were locked to Apple hardware, but the experience felt smooth, legal, and safe. Millions paid not because they had to, but because *the emotional and informational terrain had been reshaped*. The store felt trustworthy. The device felt essential. Compared to the virus-riddled chaos of Kazaa and LimeWire, iTunes felt downhill.

By the time the labels reentered the fight with digital storefronts of their own, they no longer held any high ground. Between 1999 and

2009, global recorded-music revenue fell by half. The battle wasn't lost in court—it was lost on terrain they never shaped.

## Principles in Action — Reflection Questions

1. Which element of your current environment—physical, procedural, informational, or psychological—most constrains your objective, and what modest first alteration could begin to tilt it?
2. How will you know that a shaping effort has genuinely reduced long-term friction rather than burying costs in future maintenance?
3. What sentinel indicators can warn you that an adversary is carving ground beneath your position faster than you are renewing yours?
4. When you rewrite rules or narratives, how do you protect ethical legitimacy in case power later reverses and those rules bind you?
5. After winning initial tilt, what routine inspections or cultural rituals will keep the slope intact as technology, law, and sentiment evolve?

## Sources

- A&M Records, Inc. v. Napster, Inc., 239 F.3d 1004 (9th Cir. 2001).
- Buckminster Fuller, R. 1981. *Critical Path.* New York: St. Martin's Press.
- Clark, Kenneth B., and Mamie P. Clark. 1947. "Racial Identification and Preference in Negro Children." *Journal of Negro Education* 19, no. 3: 341–50.
- Kluger, Richard. 2004. *Simple Justice: The History of Brown v. Board of Education and Black America's Struggle for Equality.* New York: Vintage.

- Knopper, Steve. 2009. *Appetite for Self-Destruction: The Spectacular Crash of the Record Industry in the Digital Age.* Brooklyn, NY: Soft Skull Press.
- Liddell Hart, B. H. 1967. *Strategy.* New York: Praeger.
- Marshall, Thurgood. 1952. "Argument Transcript: Brown v. Board of Education." U.S. Supreme Court, December 9, 1952.

# PRINCIPLE 12

## STRIKE FIRST AND SUSTAIN PRESSURE

*"Nobody ever defended anything successfully; there is only attack and attack and attack some more."* — General George S. Patton

The moment a struggle becomes unavoidable, delay is dangerous. Striking first is the decision to seize the initiative, the terrain, and the tempo before the adversary fully awakens. Yet opening blows alone rarely decide an engagement; victory accrues to the side that keeps landing follow-ups while the opponent is still blinking away surprise. This principle therefore lives in a two-step rhythm: *initiate* and *unrelentingly pursue*. An adversary that is stuck in an overwhelming defense loop cannot attack. And if they cannot attack, they cannot win. This is the heart and soul of Patton's opening quote.

That rhythm is not a reckless lunge. Preemptive action always invites danger—misreading intent, exposing one's plan, or burning irreplaceable resources. But when two conditions hold, early offense becomes overwhelmingly advantageous. First, the conflict must be judged inevitable. Whether through irreconcilable interests, locked-step escalation, or the simple ticking of a fuse, some confrontations cannot be avoided by restraint alone. Second, the actor must have done the heavy cognitive lifting of Principle 7—imagining every credible scenario and preparing branches for each. Under those circumstances, the risks of waiting dwarf the perils of moving, because initiative is the last commodity an outmatched defender can never repurchase.

Early movers exploit the adversary's cognitive lag. Surprise compresses observation time, forcing hurried orientation and guesswork. While the defender scrambles to gather facts, the aggressor is already inside the decision loop, dictating the next exchange. Tempo control follows: each fresh strike lands before the last shock dissipates, preventing consolidation, sowing doubt, and exhausting logistical bandwidth. What looks like unstoppable momentum is often just the disciplined denial of recovery windows.

Sustaining pressure, however, is neither brute repetition nor simple persistence. It is a deliberate sequencing of blows that collectively restrict freedom, multiply dilemmas, and exploit emergent weaknesses. Think of a chess grandmaster who sacrifices a pawn to open a file, then hurls a rook through that gap, only to reveal the queen already poised for mate. Each move is sequentially ordered to land just as the prior one forces realignment.

If initiative is lost, efforts should be made to regain it as quickly as possible. One of the bedrock principles of the Israeli martial art *Krav Maga* is a focus on incorporating offense into defensive motions. Doing so is an intentional effort to increase the odds of regaining initiative, thereby allowing the practitioner to transition to offense and begin pacing their attacks.

Logistics underwrite that pacing. Fuel, ammunition, cash flow, network bandwidth—these are the arteries of prolonged offense. So too are intangible stocks: public legitimacy, morale, and attention. A force that expends its magazine on the first salvo only gifts initiative back the moment it pauses to reload. Wise aggressors therefore stage reserves, rotate fresh units, and adapt tactics mid-stream—all while maintaining the appearance of uninterrupted momentum. More on this in a later chapter.

Psychology seals the advantage. Repeated shocks erode confidence faster than materiel. Neuroscientists describe the stress response as narrowing perception and degrading executive function; history illustrates the result. Defenders make unforced mistakes—patching the wrong vulnerability, redeploying forces to ghost threats, turning on one another in blame—because their mental bandwidth never recovers from the opening hit. By the time they regain composure, the board has tilted beyond repair.

Still, striking first without sufficient depth can be catastrophic. Pearl Harbor crippled the U.S. Pacific Fleet for months, yet Japan lacked the industrial stamina to exploit the blow and paid the ultimate price once American shipyards outproduced them. The lesson is not merely "hit early," but "hit early with a plan to keep swinging until the bell rings."

## Positive Example — Blitzkrieg in the West

This was the "first strike" not of World War II, but of the campaign in the West—a single, strategically unexpected blow designed to make all subsequent Allied responses futile. At dawn on May 10, 1940, German airborne units began dropping onto the Netherlands, while Panzer divisions rolled across the Ardennes—a forest the Allies believed impassable to armor. The moves looked disconnected: a paratrooper grab for Dutch airfields, diversionary thrusts into Belgium, scattered Luftwaffe raids. In reality they were phases of a tightly choreographed sequence designed to pin Allied forces north while a silent spearhead knifed through the center.

By May 12, General Heinz Guderian's XIX Panzer Corps reached the River Meuse near Sedan faster than French planners thought physically possible. Engineers laid pontoon bridges under smothering Stuka dive-bomber attacks; tanks poured across. Striking first had cracked the psychological shield of the Maginot Line without assaulting it directly.

The pressure never slackened. Rather than waiting for infantry to consolidate, Guderian drove armored columns west toward the Channel, radioing back his now-famous demand to "keep swinging"

—a phrase his staff understood literally. Each night the corps defied orders to halt, bivouacking only long enough to refuel from hastily seized French depots before racing on. On 20 May, Panzers entered Abbeville, severing the Allied front. British Expeditionary Force units, suddenly cut off, fell back on Dunkirk in disarray; evacuations began within a week. The resulting Allied chaos is immortalized in Christopher Nolan's 2017 film *Dunkirk*, which portrays confusion under relentless pressure.

Key to the sustained momentum was logistical foresight. German quartermasters shadowed the spearhead with fuel convoys rerouted from civilian stocks. Luftwaffe squadrons leapfrogged captured airfields to maintain air cover, while fresh Panzer divisions rotated into point positions whenever leading units showed fatigue. The French high command, still reeling from the Ardennes breach, attempted piecemeal counterattacks that landed against already-relocated foes, compounding panic.

By the time Paris declared an open city on June 14, Allied resistance had become a strategic rout. The campaign lasted six weeks, cost Germany fewer than 50,000 dead, and rewrote military doctrine worldwide. Critics later noted that Germany's victory depended on opponents' misjudgments, but those misjudgments were provoked, not accidental. Every Allied attempt to regroup met a timed German blow, each faster than the last, confirming Patton's maxim in real time: attack, attack, and attack some more—until the opponent forgets how to think.

## Negative Example — Microsoft Surrenders the First Move to Netscape

Few companies have ever enjoyed the market dominance Microsoft wielded in personal computing by the early 1990s. Windows 3.1 sat on more than 80% of desktops, and the firm's cash reserves dwarfed those of most competitors combined. It possessed, in UTAD terms, the raw capability described in Principle 3—enough latent power to make strategy seem optional. Yet when the internet's commercial dawn arrived, Microsoft hesitated.

In late 1994, a team of University of Illinois alumni launched Netscape Navigator, offering users a free 1.2MB download that rendered web pages in color, accepted secure payments, and—critically—ran equally well on Windows, Mac, or Unix. Navigator's first-strike impact was immediate: within twelve months it held three-quarters of global browser share and was planning an IPO valuing the start-up at nearly three billion dollars. The next strike was coming.

Microsoft's leadership, distracted by the rollout of Windows 95, initially treated the web as a curiosity. Bill Gates's famous "Internet Tidal Wave" memo did not land until May 1995—almost a year after Navigator's public release. When Internet Explorer 1.0 finally shipped that August, it arrived as a paid add-on sold through retailers, not a bundled staple. The delay yielded three disadvantages: Netscape set the de facto standards for HTML extensions, developers optimized sites for Navigator, and users formed habitual loyalty.

Realizing the danger, Microsoft pivoted hard. Windows 95 OEM agreements began requiring Internet Explorer as the default. Free upgrades rolled out quarterly. Marketing budgets exploded. Behind the scenes, engineers integrated browser hooks deep into the operating system, making removal nearly impossible. These were textbook "sustain pressure" tactics—but they were reactive, not proactive, and therefore excessively costly.

The belated offensive triggered antitrust alarms. In 1998 the U.S. Department of Justice filed suit, accusing Microsoft of wielding monopoly power to crush a smaller rival. Court testimony revealed internal emails referencing plans to "cut off Netscape's air supply." Judge Thomas Penfield Jackson's findings of fact branded Microsoft a monopolist and initially ordered a corporate breakup. Though appeals softened the remedy, the company spent over a decade under consent decrees, with compliance monitors probing product roadmaps and source code.

Microsoft's eventual victory in the browser war owed less to strategic genius than to deep pockets: it could afford to give Internet Explorer away, bankroll development, and absorb legal penalties until Netscape's parent AOL surrendered. Yet the cost was staggering —billions in legal fees, diversion of engineering talent, and a reputational dent still cited in modern antitrust debates. All of it traceable to the failure to strike first despite overwhelming capability.

Had Microsoft applied Principle 12 in conjunction with Principle 7's scenario planning, it would have recognized the inevitability of web-based competition and launched a robust, cross-platform browser ahead of Navigator. Such a move might have obviated the need for bundling warfare and spared the company the largest antitrust case of its era. Instead, Netscape's early salvo forced the giant to fight a decade-long rearguard action whose collateral damage outlived both original combatants—the very cautionary tale this principle seeks to prevent.

## Principles in Action — Reflection Questions

1. Where is conflict with a rival already unavoidable, and how could an early, well-sequenced strike tilt the timetable irreversibly?
2. Have we war-gamed every probable countermove so that sustained pressure remains feasible beyond the opening attack?
3. Which logistical resource—fuel, capital, compute cycles, political goodwill—will throttle our ability to keep hitting if we overlook it?
4. What guardrails ensure our opening blow does not violate ethical or legal limits that could later reverse public support?
5. If we hesitate, what lower-cost actor might seize the first move and force us into Microsoft-style catch-up at premium prices?

## Sources

- Boyd, John. 1986. "A Discourse on Winning and Losing." Lecture notes, Maxwell AFB, AL.
- Cusumano, Michael A., and David B. Yoffie. 1998. *Competing on Internet Time: Lessons from Netscape and Its Battle with Microsoft.* New York: Free Press.
- Frieser, Karl-Heinz. 2005. *The Blitzkrieg Legend: The 1940 Campaign in the West.* Annapolis, MD: Naval Institute Press.
- Gates, Bill. 1995. "The Internet Tidal Wave." Internal memorandum, Microsoft Corporation, May 26, 1995.
- Goenner, Alec. n.d. "The Six Pillars of Krav Maga." *Israeli Krav Maga.* http://ikmakravmaga.com/resources/the-six-pillars-of-krav-maga.html
- Guderian, Heinz. 1996. *Panzer Leader.* New York: Da Capo Press.
- Jackson, Thomas P. 1999. *Findings of Fact.* U.S. District Court for the District of Columbia, November 5, 1999.
- Keegan, John. 1989. *The Second World War.* New York: Penguin.
- Nolan, Christopher, dir. 2017. *Dunkirk.* Burbank, CA: Warner Bros. Pictures.
- Patton, George S. 1947. *War as I Knew It.* Boston: Houghton Mifflin.
- United States v. Microsoft Corp., 253 F.3d 34 (D.C. Cir. 2001).
- Watson, Richard. 1995. "Netscape's Wild Ride." *BusinessWeek,* December 11, 1995.

# PRINCIPLE 13

## TARGET WEAKNESS AND AVOID STRENGTH

*"Avoid what is strong and strike at what is weak."* — Sun Tzu

If there is one central strategy in the engagement phase that is perhaps the most ubiquitous across all adversarial domains, this is it. Conflict is almost never a duel between perfectly matched forces meeting on neutral ground. More often it resembles a sprawling chessboard where pieces differ in reach, resilience, and purpose. Armies wield tanks yet depend on fragile fuel convoys; corporations bask in market share while hiding customer resentment; bacteria mutate with dazzling speed but still rely on hospitable hosts. This principle is a two-sided coin: it is the art of avoiding an adversary's prepared defenses while striking the vulnerabilities they discounted, ignored, or sealed off with misplaced confidence.

The principle's lineage is ancient—Homer hints at it when Achilles forces Hector away from fortified Troy, and Sun Tzu states it outright—yet its logic endures because human and non-human systems grow unevenly. Power concentrates where attention and resources gather; that very concentration breeds blind spots. Supply routes thin beyond the forward edge, bureaucracies add latency, reputations ossify into predictable patterns. The stronger a fortress appears, the more likely its architects underfunded the back gate.

Strategists have refined the idea for centuries. Basil Liddell Hart called it the *indirect approach*, advising commanders to pierce the heel rather than batter the shield. John Boyd pushed further, arguing that the fastest way to disrupt an adversary's decision cycle is to target the specific weaknesses—like communication nodes or command structures—that prevent them from orienting to reality. Both converge on the same command: do not contest strength on its preferred terms; destabilize the seams that hold that strength together.

Doing so is not opportunistic flailing; it requires three deliberate disciplines.

- *Detection.* Uncovering vulnerability means mapping every dependency: logistics, morale, trust, timing, even environmental conditions. The task is part forensic science, part social anthropology—an inventory of fault lines invisible to casual observers.
- *Targeting.* Selecting the seam is an economic decision. The strike must promise high leverage, minimal self-exposure, and a ripple effect wider than the initial impact. Misdiagnose the seam and the attacker wastes effort while true power stays intact.
- *Execution.* Strength dodged is strength still present. Attacks must stay mobile, shifting pressure the moment defenses stiffen. Fixed plans ossify; adaptive sequencing thrives.

Risk shadows each phase. A weakness misread can snap shut like a beartrap. In 1941 Japan crippled Pearl Harbor—a U.S. weak point—yet overlooked America's dormant industrial colossus and paid the ultimate price. Conversely, when vulnerability is genuine and the attacker *has prepared*—imagined scenarios, stockpiled reserves, rehearsed branches—payoff is disproportionate. Morale shatters, resources hemorrhage, and the adversary may concede without a decisive battle.

For that reason, the principle pairs naturally with *Build Capability That Renders Strategy Irrelevant* (Principle 3). Overwhelming capacity grants freedom to probe for weakness without fearing a counterstrike that flips the board. Capability supplies the hammer; *targeting weakness* supplies the anvil—together they forge outcomes brute attrition could never achieve.

In practice, weakness can be as intangible as latency in a server farm or as visceral as exhausted medical staff during a pandemic. Cyberattackers bypass top-tier firewalls by instead exploiting a subcontractor's outdated authentication; insurgents avoid armored columns and instead disable bridges that feed their fuel trucks. Even nature obeys the logic: invasive kudzu smothers native plants not by withstanding winter cold, but by exploiting open sunlight gaps along highways humans maintain. We'll explore kudzu more later.

Recognizing the breadth of the principle cautions against heroic frontal charges when subtler incisions could collapse the opposition, and warns that today's weakness becomes tomorrow's reinforced bastion once exposed. Lasting success therefore lies in *habitual curiosity*, continually scanning for emerging seams while abandoning old targets before they harden into traps.

What follows are two narratives that animate the theory: Netflix's calculated assault on Blockbuster's immovable storefront empire, and a hospital's hard-learned lesson when broad-spectrum antibiotics empowered bacteria rather than patients. Both remind us that victory favors those who read the board for hidden pressure points—then strike with precision and resolve.

### Positive Example — Netflix Exploits Blockbuster's Brick-and-Mortar Anchors

I feel it may be important to point out for younger readers that Netflix started as a mail-order DVD service. This was before online film streaming existed. But before even that, we had to go to physical stores to rent videos (either VHS tapes or DVDs). Times were certainly different back then. In the late 1990s, Friday night in

America usually meant a fluorescent pilgrimage to one such physical store—Blockbuster. Six thousand locations, thirty-eight million members, and late fees that quietly added nearly a billion dollars to annual revenue formed a wall few challengers dared test. Yet those same pillars concealed fault lines.

Reed Hastings and Marc Randolph spotted them almost by accident. Renting *Apollo 13*, Hastings incurred a $40 fee and wondered aloud why plastic discs demanded stricter punctuality than library books. That irritation became data. Blockbuster's size forced it to charge penalties to offset shelf scarcity and inventory churn; customers hated the penalties. At the same time, the newly minted DVD format weighed less than a quarter pound, survived the mail, and required no rewinding. Where Blockbuster saw an inconvenient format shift, Netflix saw a postal network waiting to replace retail rental space.

Netflix's first incision eliminated late fees. A flat monthly subscription let members queue titles online and return discs whenever the red-and-white envelope found its way back to a blue mailbox. In doing so, they created positive incentive for customers to return movies, in contrast to Blockbuster's punitive late fees. The move turned Blockbuster's cash cow into a liability: dropping fees would vaporize revenue, yet keeping them magnified resentment. While executives debated, Netflix probed deeper.

The second strike arrived in 2000 with *Cinematch*, an algorithm that recommended films based on user ratings. This seems commonplace now, but it was *revolutionary* at the time. Personalization required data, which Blockbuster's stores could not (or at least did not) capture at scale. By harvesting viewing habits, Netflix fortified customer loyalty and discovered hidden demand for niche titles unsuited to limited shelf space—another seam Blockbuster could not patch without overhauling procurement and display logistics.

Then came streaming. By 2007, broadband saturated half of U.S. households. Hastings launched Watch Now, accepting grainy resolution and limited catalog in exchange for zero postage and instantaneous gratification. Blockbuster hesitated, fearing that cannibalizing foot traffic would collapse its real estate-heavy model before an online replacement matured. The hesitation was fatal. This marked a critical turning point: the moment Netflix's business model had fully pivoted from physical media to digital streaming—a transition Blockbuster was unable to make. In 2010, Blockbuster filed for Chapter 11.

Netflix's campaign was a masterclass in all three disciplines. *Detection* began with Reed Hastings's personal frustration over a late fee, revealing customer resentment as a key vulnerability. *Targeting* was precise: first the late-fee model, then the limitations of physical shelf space with the Cinematch algorithm. *Execution* was a patient, sequential attack, pivoting from mail-order DVDs to streaming only when the technological and market conditions were right, ensuring Blockbuster was always reacting to the last move.

At no point did Netflix outmuscle Blockbuster's capital or brand. Instead, it targeted Blockbuster's vulnerabilities by sequencing pressure on late fees, shelf scarcity, data blindness, and its real estate overhead—each strike timed to exploit the moment Blockbuster tried to pivot. Netflix won by treating Blockbuster's greatest strengths—its physical stores and late fee revenue—as vulnerabilities to be systematically dismantled.

### Negative Example — Broad-Spectrum Antibiotic Overuse Breeds a Super-Bug

By late 2008, Gregorio Marañón University Hospital in Madrid was facing a problem of its own making. Known for aggressive infection control, the hospital's intensive-care units had begun to incubate an adversary that evolved precisely because of how it was attacked. As sepsis alerts rose, clinicians defaulted to broad-spectrum antibiotics —blanketing patients with pharmaceutical firepower in hopes of halting infection before blood cultures could identify the exact cause. The intention was speed. The outcome was escalation.

At first, the approach worked. But by February 2009, three patients tested positive for Methicillin-resistant *Staphylococcus aureus* (MRSA). By April, the count had climbed to thirteen, spanning multiple wards. Genome sequencing traced the outbreak to a single lineage bearing the *cfr* gene—conferring resistance to linezolid, one of the last pharmacological lines of defense.

Why had the hospital's best weapons failed? Because the strategy struck directly at the pathogen's *strength*: rapid mutation under selective pressure. Each broad-spectrum dose was an evolutionary sorting mechanism, clearing the field of weak microbes and promoting survivors built to resist. The more aggressive the assault, the faster the pathogen adapted. In Clausewitzian terms, the hospital had attacked the enemy where it was strongest—and strengthened it in the process.

Meanwhile, MRSA's true vulnerability—its reliance on physical surfaces and human carriers—went unaddressed. Biofilms accumulated in ventilator tubing. Cleaning protocols emphasized public spaces but missed critical contact points. Visitors drifted between rooms without consistent hand hygiene. The hospital targeted the biochemical threat inside the body while leaving untouched the *operational terrain* that enabled its spread.

When mortality reached six of eighteen infected patients, administrators convened an emergency stewardship board. This time, the strategy changed. Linezolid use dropped 90% within a quarter. Broad-spectrum policies were replaced with narrow-spectrum regimens tailored to identified strains. Infection-control nurses implemented real-time surveillance of ventilator equipment and enforced hygiene protocols with surgical precision. The pivot was decisive. Within six months, the outbreak curve flattened.

The hospital's failure was a breakdown in *detection* and *targeting*. Clinicians correctly detected the presence of infection but failed to detect its weakness: reliance on physical transmission. They thereby incorrectly targeted the pathogen's strongest evolutionary trait—its ability to resist drugs. This underscores a timeless strategic error: frontal assaults on an adversary's primary advantage often strengthen it. The successful *execution* of a new strategy, targeting hygiene protocols, proved the lesson: victory comes not by overpowering a rival, but by striking where adaptation is least possible and collapse is most likely.

## Principles in Action — Reflection Questions

1. Which fees, frictions, or legacy policies in our rival's model generate simmering resentment we could flip into loyalty?
2. How frequently do we revisit our adversary map to ensure yesterday's weakness has not hardened into today's bastion?
3. In what ways might our strongest capability—supply chain, culture, trademark—double as rigidity a nimble challenger could exploit?
4. What rapid-feedback loops (customer surveys, threat intel, lab cultures) can alert us to emerging seams before opponents notice them?
5. Could our strike inadvertently strengthen the opponent— evolving them into a more dangerous version—and how will we mitigate that risk?

## Sources

- Boyd, John. 1986. *Patterns of Conflict*. Briefing slides.
- Clausewitz, Carl von. 1984. *On War*. Translated by Michael Howard and Peter Paret. Princeton, NJ: Princeton University Press.
- European Centre for Disease Prevention and Control. 2012. *Healthcare-Associated Infections: Annual Epidemiological Report 2012*. Stockholm: ECDC.
- Fritz, Martina, et al. 2011. "Linezolid-Resistant MRSA Outbreak in Intensive-Care Units, Madrid, 2008–09." *Clinical Infectious Diseases* 52, no. 5: 676–84. https://doi.org/10.1093/cid/ciq202

- Homer. 1990. *The Iliad.* Translated by Robert Fagles. New York: Penguin Classics.
- Huang, Jeff. 2014. "How Netflix Beat Blockbuster: A Business Model Innovation Study." *Harvard Business Review,* January 14, 2014.
- Keating, Gina. 2012. *Netflixed: The Epic Battle for America's Eyeballs.* New York: Portfolio/Penguin.
- Liddell Hart, B. H. 1967. *Strategy.* 2nd ed. New York: Praeger.
- Randolph, Marc. 2019. *That Will Never Work: The Birth of Netflix and the Amazing Life of an Idea.* New York: Little, Brown.
- Sun Tzu. 1963. *The Art of War.* Translated by Samuel B. Griffith. Oxford: Oxford University Press.
- Surowiecki, James. 2010. "Last Blues for Blockbuster." *The New Yorker,* October 18, 2010.
- U.S. Census Bureau. n.d. "Retail Video Store Revenues, 1990–2010."

# PRINCIPLE 14

## REDUCE YOUR EXPOSURE

*"Suppose we were ... a thing intangible, invulnerable, without front or back, drifting about like a gas."* — T. E. Lawrence

The previous chapter explained how to win by targeting an adversary's weakness. This principle is the logical counter: how to win by making your own weaknesses difficult to target. The goal is to frustrate your opponent's ability to strike with effect by deliberately shaping what you expose to them.

This isn't passive defense; it's an active strategy to change the economics of the engagement. By making yourself a more expensive, time-consuming, and uncertain target, you shift the burden of effort to the other side. This is achieved by systematically reducing what's available to hit, obscuring what remains, spreading value across locations, dividing systems internally, and refusing to stay in one place long enough to be predictable.

Five strategic moves form the foundation of this principle:

1. **Minimize** – Reduce what's available to target in the first place
2. **Conceal** – Obscure the visibility of what remains

3. **Disperse** – Avoid putting too much value in one location
4. **Segment** – Limit how far damage can spread
5. **Move** – Keep shifting to stay ahead of the attacker's map

Each element works on its own. Together, they shift the burden of effort to your adversary—and tilt the conflict in your favor.

The same five moves apply to groups. A wolf pack uses foliage to *conceal* its approach. A military patrol will *minimize* its time spent in a dangerous open area. A herd of zebra will *disperse* to ensure a predator cannot attack the whole group at once. In a pandemic, health officials use *segmentation* by cohorting patients to prevent a single outbreak from spreading through an entire hospital. And a flock of starlings will *move* in a murmuration, their coordinated, unpredictable flight making it nearly impossible for a falcon to target any single bird.

## Minimize: Reduce What's There to Hit

The most reliable way to reduce your exposure is to remove what doesn't need to be exposed in the first place.

Martial artists who focus on striking are taught to turn their bodies sideways. That posture shrinks the size of the attack zone and reduces the number of available targets. A square stance gives opponents too much to work with. A narrow one makes clean hits harder to land. Similarly, architects of modern stealth warships use sloped hulls and low profiles above the waterline. This shrinks the ship's radar cross-section, reducing the area an enemy missile has to target. The same logic applies across domains: if something doesn't need to be exposed, it shouldn't be left available.

Minimization means reducing the number of systems, services, roles, permissions, and surface area that can be detected or used against you. It's not hiding—it's deletion. Eliminating low-value endpoints, revoking stale access credentials, consolidating processes, retiring unmaintained assets—these aren't technical cleanups, they're strategic denials. What isn't present can't be exploited. What doesn't run can't be hijacked. What no longer exists doesn't need to be defended.

The more you remove, the less there is to attack. The smaller your presence, the harder it becomes to gain traction against it.

## Conceal: Obscure What Remains

Once you've minimized, the next move is to limit what can be seen or interpreted. Concealment doesn't eliminate a target—it masks it. The goal is to disrupt the attacker's ability to identify, track, and prioritize assets.

In combat, concealment means camouflage, smoke, or misdirection. In infrastructure, it means encrypted communications, non-descriptive system labels, obscured architectures, and noise-injected telemetry. Concealment is what frustrates targeting, what prevents fingerprinting, what forces an adversary to guess.

The less the opponent can see, the more likely they are to waste time. They may follow false signals. They may strike irrelevant assets. They may delay action while they recheck their assumptions. Every layer of concealment introduces uncertainty, and uncertainty slows the fight.

## Disperse: Avoid Critical Concentration

Efficiency is often the enemy of resilience. When everything is centralized—one database, one warehouse, one lead decision-maker —you're faster on a good day and vulnerable on a bad one.

Dispersal solves that. It spreads value across locations, systems, or people. It creates independent nodes that cannot all be struck at once. This doesn't just reduce the risk of failure. It multiplies the attacker's workload.

When infrastructure is dispersed, adversaries must choose their targets carefully. When decision-making is distributed, disruption requires broader coordination. When storage, compute, and backup are located in different jurisdictions, attackers face legal and logistical delays. Dispersal breaks momentum. It ensures that even a good hit can't do too much at once.

Some centralization is inevitable. But too much turns you into a single point of failure. Dispersal forces the adversary to solve a harder problem—and often, to solve it repeatedly.

## Segment: Prevent Damage from Spreading

No system is unbreakable. Something will eventually be hit. Segmentation makes sure it doesn't spread.

This means designing boundaries within your organization, your infrastructure, your data, and your process flow. It's about isolation—building compartments that fail independently, rather than cascading into system-wide failure.

In security, this means sandboxing execution environments, limiting lateral movement between services, and enforcing access controls that restrict what any one actor can see or change. In physical systems, this looks like firebreaks, bulkheads, or modular supply chains. In organizations, it means keeping responsibilities, privileges, and visibility tightly scoped to need.

Segmentation doesn't stop the first strike. It ensures there's no second wave. When attackers are forced to start over each time, their campaign slows down or collapses entirely.

## Move: Stay in Motion to Stay Ahead

Everything up to this point makes you harder to hit. But staying still —even well-hidden and well-structured—lets adversaries map you over time. The final layer of exposure reduction is movement.

To be dynamic is to expire tokens before they can be reused. To rotate infrastructure faster than it can be fingerprinted. To shift IP addresses, change DNS records, and rebuild containers automatically before they grow stale. It's the practice of creating a moving target— one that's not just concealed but outdated by the time someone reaches for it.

In martial arts, this is footwork—changing angle, range, timing, rhythm. In cybersecurity, it's ephemeral workloads. In operations, it's alternating locations, shifting routes, and rotating responsibilities. The more often you change, the more fragile your opponent's intelligence becomes. Their map stops matching reality.

Movement forces the attacker to fight the environment, not just you.

## Friction Asymmetry: Why It Works

These five moves work because they shift effort to the other side.

You spend modest resources to shrink, hide, spread, divide, and shift your presence. Your adversary must now spend outsized resources to locate, analyze, synchronize, penetrate, and pursue it. This imbalance is called friction asymmetry—and it's one of the defender's best assets.

Attacks slow down. Mistakes multiply. Confidence drops. Each step forward costs more than the last.

Reducing your exposure doesn't mean you can't be touched. It means you stop being a clean target—and start becoming a costly one.

## Bottom Line

Exposure invites exploitation. When you're visible, centralized, over-connected, and stationary, you're easy to plan against—and rewarding to strike.

Reduce what exists. Obscure what remains. Avoid piling value into one place. Wall off internal connections. Refuse to stay put.

*Make yourself inconvenient. Make them chase. And make every shot they take cost them more than it costs you to absorb it.*

## Positive Example — Cyber Exposure Reduction Before and After Breach

Some organizations reduce exposure by design. Others reduce it after the damage is done. Google and Capital One followed different time-lines, but both ended up proving the same principle: when attackers struggle to find, reach, and hit valuable targets, they often stop trying.

In 2009, Google was hit by a highly coordinated cyber operation known as *Operation Aurora*. The attackers entered through remote-access infrastructure and moved laterally to access internal systems, including source code repositories. In the aftermath, Google aban-doned traditional perimeter defense and launched a new internal architecture called *BeyondCorp*—a forerunner of what would later be called Zero Trust.

Every layer of the redesign reflected exposure reduction in action.

Google *Minimized* by collapsing broad internal access and reducing the number of exposed services and entry points. Legacy infrastructure that once allowed lateral movement across the network was dismantled. Access to applications was rebuilt around tightly scoped, identity-bound requests. Systems that no longer met expo-sure standards were shut down, replaced, or isolated. The number of targets available to an attacker dropped sharply—not just through policy, but through structural reduction.

They *Concealed* their internal infrastructure behind identity-aware proxies. Even if attackers scanned internal systems, they couldn't see what applications were running or how they were configured.

They *Dispersed* authentication and access enforcement across a global edge network. There was no single choke point to attack and no predictable path for intrusion.

They *Segmented* aggressively. Each application lived in its own protected zone. Identity credentials expired frequently, and policies refreshed with every request. If an attacker gained a foothold, there was nowhere to move without triggering new access checks.

Most importantly, they *Moved*. Endpoints rotated. Tokens turned over constantly. Devices had to reverify. Even if an attacker got in, they couldn't stay in for long.

Google didn't just defend itself—it reshaped the cost structure of attacking it. As security researchers often note, well-defended Zero Trust organizations are frequently bypassed by attackers who deem them too costly or time-consuming to pursue. Exposure reduction doesn't guarantee immunity, but it encourages the adversary to look elsewhere.

Capital One learned the same lesson—just later.

In 2019, a misconfigured web application firewall exposed a Server-Side Request Forgery vulnerability. An attacker exploited it to access internal metadata and escalate into S3 storage buckets holding the personal information of approximately 106 million individuals, which included credit application data. The breach revealed a system with partial defenses but too much exposure—one layer missing, one check skipped, one assumption left in place.

After the incident, Capital One rebuilt with exposure reduction in mind.

They *Minimized* attack surface by aggressively removing outdated services and performing daily DNS sweeps for forgotten subdomains.

They *Concealed* application access behind stricter identity layers, ensuring internal resources weren't easily scanned or resolved.

They *Dispersed* workloads across multiple environments with automated asset inventory updates, making it harder to predict where anything was running at a given time.

They *Segmented* roles, permissions, and traffic flows more tightly. Infrastructure-as-code deployments now include automated checks that block any increase in attack surface.

They *Moved* toward automation-driven change. Containers, routes, and identities update constantly. Static targets became temporary ones. By the time an attacker maps something, it's already been replaced.

The breach cost Capital One hundreds of millions of dollars. But the rebuilt model is now treated as a benchmark for secure architecture across the banking industry. Reducing exposure didn't just improve security—it became reputational capital.

Whether before a serious attack or after one, the outcome is the same: less to see, less to strike, less to lose.

## Negative Example — Passenger Pigeons: When Abundance Becomes Bullseye

Passenger pigeons once filled North American skies in flocks numbering in the billions. Their evolutionary defense strategy was scale. By traveling and nesting in massive groups, they overwhelmed predators. Any individual bird had a high chance of survival because the odds of being singled out were vanishingly low. The more of them there were, the safer each one became.

But industrialization inverted that logic.

Telegraph lines allowed hunters to broadcast sightings across long distances. Railroads and ice-cooled storage turned local kills into national shipments. Shotguns—already ideal for hunting birds in flight—could bring down dozens of pigeons with a single blast when aimed into a dense flock. Hunters didn't need to track individual birds—they followed wires, timetables, and smoke.

The pigeons' strategy, once effective, collapsed under the weight of its own predictability. They had violated every element of exposure reduction.

There was no *Minimization*. Their survival depended on overwhelming numbers and visibility. They existed in high density, and they needed to be seen to be together.

There was no *Concealment*. Flocks were obvious, loud, and inescapable once spotted. Nesting colonies stretched for miles and drew attention from hundreds of miles away.

There was no *Dispersion*. Breeding required synchronized mass gatherings. Their survival model was centralized by design. There were no backup colonies, no distributed fallback populations.

There was no *Segmentation*. Once a nesting ground was located, hunters could disrupt an entire reproductive cycle. There were no buffers or boundaries—just one vulnerable cluster after another.

And there was no *Movement* discipline. The pigeons followed the same migration routes each year. They returned to the same nesting grounds at predictable times. They didn't change patterns fast enough to outrun the accelerating systems built to exploit them.

As their numbers declined, hunters mistook the thinning flocks as a sign that larger ones must exist elsewhere. So they hunted harder. Each year's smaller gatherings were found faster, shot more efficiently, and shipped farther. By 1901, the wild population was functionally extinct. Martha, the last known passenger pigeon, died in captivity in 1914.

The pigeons didn't fall to a new predator. They fell to a modern targeting system that turned visibility and centralization into liabilities. What once protected them became the pattern that doomed them.

The same dynamic appears today in digital infrastructure. Massive data lakes guarded by a single identity provider. Application clusters managed through one orchestration plane. Monocultures of software, network, or logistics. As with the pigeons, these systems work well—until the tools to target them scale faster than the defenders can adapt.

Exposure reduction isn't a luxury. It's a form of resilience. And when it's missing, even the biggest population, company, or system can collapse in full view—simply because it made itself too easy to hit.

## Principles in Action — Reflection Questions

1. Which components of our operation remain exposed out of habit rather than necessity, and what would it cost to retire or mask them?
2. How frequently do we rotate or randomize unavoidable

surfaces—API gateways, shift schedules, shipping lanes—
so yesterday's reconnaissance expires uselessly?

3. Where have we accepted monoculture efficiency (single
   suppliers, centralized identity, giant data buckets) that a
   modest disruption could paralyze?

4. Does our patch cadence and key-rotation rhythm force
   attackers to restart reconnaissance faster than they can
   automate it?

5. How do we reassure partners and regulators that limited
   visibility is *strategic prudence* rather than negligence or
   secrecy?

## Sources

- Bromiley, Matt. 2023. "Know Thyself, Know Thy Enemy: A
  Proactive Approach to External Attack Surface
  Management." *SANS Institute Whitepaper.*
- Google. 2014. *BeyondCorp: A New Approach to Enterprise
  Security.* White paper. Mountain View, CA: Google.
- Greenberg, Andy. 2020. "The Untold Story of the Capital
  One Hack." *Wired,* December 15, 2020.
- Greenberg, Joel. 2014. *A Feathered River Across the Sky: The
  Passenger Pigeon's Flight to Extinction.* New York:
  Bloomsbury USA.
- Kindervag, John. 2010. *Build Security into Your Network's
  DNA: The Zero Trust Network Architecture.* Forrester
  Research.
- Lawrence, T. E. 1926. *Seven Pillars of Wisdom.* London:
  Jonathan Cape.
- National Institute of Standards and Technology. 2020. *SP
  800-207: Zero Trust Architecture.* Gaithersburg, MD: NIST.
- Schorger, A. W. 1955. *The Passenger Pigeon: Its Natural
  History and Extinction.* Madison: University of Wisconsin
  Press.

- U.S. House Committee on Oversight and Reform. 2022. *Hearing on the Capital One Data Breach.* May 13, 2022.

# PRINCIPLE 15
## EXPLOIT THE ENVIRONMENT

*"How to make the best of both strong and weak—that is a question involving the proper use of ground."* — Sun Tzu

The environment of conflict—its *physical, procedural, informational,* and *psychological* dimensions—is never just a backdrop. It actively shapes what each side can do, when, and how. *Exploit the Environment* is the art of using those existing features to constrain your adversary, reduce their options, and channel them into predictable moves. Unlike *Shape the Environment* (Principle 11), which is about preparing terrain in advance, exploitation means taking what is already present—often immutable—and turning it ruthlessly to your advantage.

Warren Buffett, the longtime head of Berkshire Hathaway, popularized the phrase "economic moat" as an investor's shorthand for what makes one company more durable than another. Crucially, he argued that this durability often came not from the company itself but from the *environment* around it. Just as a medieval castle's moat forced attackers to pay a higher price to assault its walls, a firm with the right environmental defenses could force competitors to spend heavily just to compete. For Buffett, identifying such moats was the

essence of investing: look for businesses whose surroundings—
market structures, customer habits, or regulatory conditions—
shielded them in ways rivals could not easily breach.

These moats take many forms. A powerful brand is a feature of
the *psychological terrain*, creating trust and loyalty before a transaction
even begins. High switching costs or exclusive patents are features of
the *procedural terrain*, limiting competitors' ability to pry customers
away. The point is not that the company builds a new defense each
day, but that the environment itself provides the defense. Like water
around a fortress, these moats turn structural conditions into protection, allowing the company to conserve strength while forcing rivals
to spend disproportionately just to reach the fight.

The four terrains of conflict each offer exploitable opportunities:

**Physical Terrain** — Geography, climate, and built infrastructure
can be used to amplify your strengths or degrade the enemy's. A
northern football team that schedules a late-season home game
against a southern rival is exploiting weather conditions as a weapon.

**Procedural Terrain** — Rules, laws, and protocols can be turned
into choke points. A trial lawyer excludes key testimony by invoking a
rule of evidence, narrowing the opposition's room to maneuver.

**Informational Terrain** — Narratives and signals already circulating can be leveraged to magnify impact. A campaign amplifies a
rival's out-of-context quote that dovetails with a popular negative
storyline, letting the environment itself do the work.

**Psychological Terrain** — Cognitive biases and cultural assumptions can be manipulated. A negotiator hosts talks in their own
imposing office, exploiting setting and hierarchy to induce deference
without a word.

∾

THE PURPOSE OF EXPLOITATION IS TO MAKE THE ADVERSARY'S PATH
narrower, harder, and more predictable. A column funneled into a
mountain pass, a corporation bound by precedent in court, or a polit-

ical opponent boxed in by public opinion each faces constrained choices. Predictability breeds vulnerability.

This principle often favors the weaker side. Guerrilla fighters use jungles, mountains, or cities to cancel out an adversary's advantages in armor and aircraft. A startup exploits regulatory carve-outs to outmaneuver incumbents. By turning environmental features into weapons, the weaker actor multiplies its force and erodes the other side's.

But exploitation punishes ignorance as quickly as it rewards mastery. You cannot weaponize an environment you do not understand. Applying this principle requires the rigor of a geographer mapping terrain, a lawyer dissecting procedural rules, and a sociologist tracing cultural and informational currents. Failures in that intelligence function lead to disaster: armies swallowed by swamps they failed to chart, lawsuits collapsing when a precedent has been overturned, marketing campaigns backfiring against deeply held cultural beliefs. To exploit the environment, you must first become its most devoted student.

### Positive Example: The Battle of Thermopylae

Surely, you knew this example was coming. The Battle of Thermopylae in 480 BC is the timeless and quintessential positive example of a vastly outnumbered force exploiting the physical environment to neutralize a superior adversary. The adversarial dynamic was one of extreme asymmetry: a small coalition of Greek city-states, led by a core of 300 Spartans under King Leonidas, faced the massive invading army of the Persian Empire, estimated by modern historians to be in the hundreds of thousands.

The Greek strategy was based entirely and brilliantly on exploiting the environment. On an open field, their small force would have been enveloped and annihilated in minutes. Recognizing this, they chose not to fight on open ground. Instead, they made their stand at the coastal pass of Thermopylae, known as the "Hot Gates."

This location was a natural chokepoint, a narrow passage with impassable mountains on one side and the sea on the other.

This deliberate choice of terrain weaponized the geography itself. The narrowness of the pass rendered the Persians' primary strength —their overwhelming numbers—completely irrelevant. They could only bring a fraction of their army to bear against the Greek front at any one time, forcing them into a brutal, head-on fight. This, in turn, amplified the Greeks' primary strength: the superior training, discipline, and heavy armor of the Spartan hoplites. The environment forced the Persians to fight the exact type of battle the Spartans were best equipped to win. For three days, the small Greek force used this environmental chokepoint to hold off the massive Persian army, inflicting devastating casualties. While the Greeks were eventually defeated after being outflanked via a hidden mountain path (a failure to secure all aspects of the environment), their stand at the pass remains the ultimate historical testament to how a small, determined force can exploit the terrain to impose its will on a much larger one.

## Negative Example: The Spread of the Emerald Ash Borer

The catastrophic spread of the invasive Emerald Ash Borer (EAB) across North America is a powerful negative example of a failure to exploit the environment, drawn from the domain of cross-species conflict. In this case, it is a story of humans failing to use the environmental features they themselves control to contain a devastating adversary.

The adversarial dynamic here is between humans, seeking to preserve North America's ash tree populations, and the EAB, an adaptive and highly destructive insect from Asia. The beetle's larvae feed on the inner bark of ash trees, cutting off their flow of water and nutrients and killing them within a few years. While the beetle can fly, its natural spread is relatively slow. The primary vector for its rapid, long-distance invasion is the human transportation of infested firewood.

The failure is one of environmental exploitation. The man-made environment of North America is filled with perfect chokepoints and guidance zones that could have been used to severely constrain the beetle's movement. These include highways, state and international borders, agricultural inspection stations, and the clearly defined entrances to state and national parks. A strategy based on this principle would have involved the early and rigorous exploitation of these chokepoints through comprehensive and strictly enforced quarantines and bans on the movement of firewood.

Instead, for years the response was fragmented and insufficient. Public awareness campaigns were slow to take hold, and regulations on firewood transport were inconsistent and poorly enforced across jurisdictions. This failure to use the existing, human-controlled environment to trap and contain the adversary allowed the EAB to spread uncontrollably. Our own transportation networks—the very environmental features that could have been weaponized against the beetle —instead became the primary weapon *used by the beetle* to expand its territory at a speed many times faster than it could have achieved naturally. It is a stark example of how failing to exploit the chokepoints in your own environment can lead to a complete and devastating strategic loss.

## Principles in Action — Reflection Questions

1. What is the "terrain" of our current conflict? Have we fully mapped its physical, procedural, informational, and psychological features?
2. What is the single most powerful environmental feature— a chokepoint, a rule, a public belief—that we are not currently using to our advantage?
3. How can we use the environment to force our adversary into a predictable course of action, thereby reducing their options and making them easier to counter?
4. What is our adversary's relationship with the environment? Are they more or less adapted to it than we

are, and how can we exploit that difference (like the
football team in the snow)?

5. If we were to look at our own operational environment,
what is the key feature that our adversary is currently
exploiting against us, and what is our plan to neutralize
that advantage?

## Sources

- Buffett, Warren. 2007. *Annual Letter to Berkshire Hathaway Shareholders.* Omaha, NE: Berkshire Hathaway Inc.
- Herodotus. 1996. *The Histories.* Translated by G. Rawlinson. London: Penguin Classics.
- Holland, Tom. 2007. *Persian Fire: The First World Empire and the Battle for the West.* New York: Anchor Books.
- Poland, Therese M., and Deborah G. McCullough. 2006. "Emerald Ash Borer: Invasion of the Urban Forest and the Threat to North America's Ash Resource." *Journal of Forestry* 104, no. 3: 118–24. https://doi.org/10.1093/jof/104.3.118
- Sun Tzu. 1963. *The Art of War.* Translated by Samuel B. Griffith. Oxford: Oxford University Press.
- United States Department of Agriculture (USDA). 2021. *Emerald Ash Borer Program Manual.* Animal and Plant Health Inspection Service.

# PRINCIPLE 16
## TIME YOUR ACTIONS STRATEGICALLY

*"There is a tide in the affairs of men, which, taken at the flood, leads on to fortune."* — William Shakespeare

Most people understand the stakes of timing long before they ever study strategy. Think about romance. A first kiss that arrives when tension has built and both people are leaning in—literally or emotionally—can seal the moment. The same kiss, delivered too soon or without a shared signal, feels abrupt or out of sync. Say "I love you" on the second date and it might trigger alarm. Wait too long, and it may stir doubt or disappointment. The actions don't change—but the moment they land determines how they're received.

Strategic conflict operates under the same logic. Opportunities surface, crest, and vanish, and the value of a decision is often measured less by its content than by its alignment with those waves. The Greeks distinguished between *chronos*—sequential, measurable time—and *kairos*—"the decisive moment." Great campaigns harmonize the two: they build sustainable rhythms while remaining alert for the brief instant when action has disproportionate effect.

That harmony rests on three intertwined habits. They don't

require brilliance or aggression—just disciplined timing, practiced over time.

**Preparation.** Ships must be caulked and crews drilled before the tide surges. Preparation is unglamorous, easily postponed, and often the decisive half of timing. Pfizer's rapid COVID-19 vaccine rollout succeeded not because it rushed, but because it was ready. Its partner, BioNTech, was ahead of the game. Its founders had spent two decades advancing mRNA science before the company was formed. When the opportunity struck, they didn't start from scratch—they scaled what already existed. By contrast, Argentina's 1982 Falklands gamble was launched with harsh seas and winter weather on the horizon and logistical hubs under-stocked. The plan itself was bold; the timing was a key factor that made it brittle.

**Sensing.** A strategist must recognize the faint signal that separates noise from opportunity. In currency markets, that signal might be a central-bank statement. In social movements, a viral image or video. On a battlefield, a sudden pause in artillery fire that reveals ammunition depletion. Missing the signal can be fatal. When Union scouts captured Confederate General Robert E. Lee's Special Order 191 in 1862, it revealed his forces were dangerously split. But General George McClellan hesitated for nearly eighteen hours before moving. That delay gave Lee just enough time to regroup at Antietam—turning what could have been a decisive blow into a costly stalemate.

**Tempo.** Once the window opens, rhythm determines whether you control it—or get crushed in it. Some moments require speed: paratroopers seizing a bridge must be followed by armor before defenders regroup. Others reward restraint: Apple's long pause between iPhone prototype and product launch generated massive consumer anticipation. Strategic tempo means knowing when to push and when to hold, when to accelerate and when to slow—so your adversary never sets the pace.

When preparation, sensing, and tempo align, timing stops being guesswork. What others experience as luck becomes something closer to inevitability.

**Positive Example — George Soros and Black Wednesday**

George Soros's attack on the British pound is often reduced to a headline: *"the man who broke the Bank of England."* But beneath the drama was a meticulous application of timing discipline.

The backdrop was the European Exchange Rate Mechanism (ERM), a precursor to the euro designed to keep participating currencies within fixed exchange bands. In the early 1990s, Britain was bound to keep the pound trading within a narrow range relative to the German mark. But economic fundamentals made that promise hard to keep: Britain was struggling with inflation, stagnant growth, and political fragility, while Germany—still digesting the cost of reunification—was facing upward pressure on interest rates.

Soros believed the pound was overvalued and that Britain wouldn't be able to hold its ERM commitment if pressure mounted. He saw an opportunity and began preparing months in advance. Starting in April 1992, his Quantum Fund analysts quietly built positions, borrowing pounds through forward contracts and accumulating massive short exposure—dispersed across tranches small enough to avoid notice. At the same time, they modeled how British policymakers might react: how quickly reserves might drain, how far interest rates could be pushed, how long coalition politics might delay intervention. When September arrived, they were not guessing. They were ready.

Then came the signal. On September 15, the president of Germany's Bundesbank gave a cautious statement that signaled a shift: Germany would not raise interest rates in coordination with the rest of the ERM. To most readers, it was a vague comment. To Quantum Fund analysts, it was a green light.

At dawn the next day—September 16, 1992—Soros struck. His fund executed wave after wave of sterling-for-mark swaps, spaced just far enough apart to keep markets open but too close for officials to regroup. As the pound weakened, other funds joined in. Britain raised interest rates from 10 to 12%, then to 15% in a matter of hours—but Soros had left himself room to re-enter at each level, applying pressure without overextending.

By nightfall, the British government conceded. The pound fell out of the ERM. The Treasury had spent over £3 billion trying to defend it. Soros reportedly made $1 billion in profit.

But the real lesson wasn't in the payout—it was in the timing. Months of silent preparation. Seconds of precise execution. A rhythm of escalation that kept intervention forces reactive. And finally, a controlled unwind of his positions—gradual enough to avoid regulatory backlash or a chaotic rebound. The action itself wasn't new. Many speculators had bet against currencies. Soros simply picked the moment when the move would matter most.

Soros's victory was a perfect execution of all three disciplines of timing. His *Preparation* was the months-long, quiet accumulation of a massive short position. His Sensing of *kairos*, the decisive moment, was triggered by the German Bundesbank's subtle statement. Finally, his control of *Tempo* was masterful; he applied pressure in waves, forcing the Bank of England to react to his rhythm, never allowing it to regain the initiative.

### Negative Example — Kodak's Digital Delay

Kodak's fall is sometimes cast as stubbornness against innovation, yet the archival record shows the company *invented* most of the technologies that later destroyed it. The fatal flaw was mistiming every pivot.

*Too soon, then too late.* Steve Sasson's 1975 eight-pound digital prototype amazed executives but threatened a film empire generating 80% of corporate profit. Management filed patents and returned to emulsion chemistry. By the late 1980s, as Sony's Mavica gained buzz, Kodak finally funded digital R&D—but directed engineers to preserve print as the central value proposition, assuming consumers still craved physical photos. They set a leisurely three-year product cycle—glacial by silicon standards—hoping to graft new tech onto the old revenue moat.

*Window narrows.* Between 1999 and 2003, sensor resolution leapt from one to five megapixels while storage prices collapsed. Camera-phone prototypes from Sharp and Nokia hinted that the entire capture-to-share chain could bypass printing. Kodak executives, alarmed but encumbered by film-plant obligations and retailer alliances, accelerated launches—but now faced a two-front war: nimble Japanese camera makers on one flank and telecom giants on the other. Capital scattered across factories, online galleries, kiosk networks, and belated CMOS fabrication bets.

*Tempo mismatch.* In 2007 Apple's iPhone reorganized the battle-field. Annual smartphone refresh cycles crushed Kodak's multi-year roadmaps; app ecosystems replaced desk-docked EasyShare software with frictionless cloud uploads. Kodak responded by cutting 27,000 jobs to free cash for sensor plants—precisely as the market pivoted from dedicated cameras to pocket hybrids. The company's tempo lagged at every escalation, burning reserves in sprints that ended just as the race route changed. In January 2012, the once-powerful Kodak filed for bankruptcy, its patents auctioned to the very smartphone manufacturers it once dismissed.

Kodak's failure was a breakdown across all three disciplines. Their *Preparation* was incomplete; they invented the technology but failed to build a business model for it. Their *Sensing* was fatally slow, consistently underestimating the speed of the digital shift until it was too late. Most critically, their *Tempo* was a complete mismatch for the environment; they operated on glacial, multi-year product cycles while the market was accelerating to an annual, then near-instanta-neous, pace of innovation.

*Strategic lesson.* Innovation without temporal alignment breeds strategic whiplash. Kodak sensed the tide early but refused to sail; when it finally launched, the flood had turned and carried rivals downstream on momentum Kodak's late oars could not match.

## Principles in Action — Reflection Questions

1. Which external signals—legislative drafts, competitor downgrades, climate anomalies—compose our early-warning grid for kairos, and who is accountable for watching each beacon?

2. Where are we stockpiling superfluous inventory, cash, or
   goodwill that could decay before a window opens, and
   how might we redeploy it into *clock-neutral* assets like
   training or modular tooling?

3. Does our project cadence contain *built-in slippage buffers* so
   that unforeseen opportunity can be seized without
   derailing core operations?

4. What specific metric alerts leadership that the window is
   closing—market-share inflection, troop fatigue rate,
   bandwidth saturation—and triggers immediate tempo
   shift or exit?

5. How do we rehearse tempo changes (accelerations,
   decelerations, oscillations) so the organization's cultural
   muscle memory responds fluidly rather than jolting from
   inertia?

## Sources

- Bank of England. 1993. *Report on Withdrawal from the
  Exchange Rate Mechanism.* London: HMSO.
- Chancellor, Edward. 1999. *Devil Take the Hindmost: A
  History of Financial Speculation.* New York: Farrar, Straus
  and Giroux.
- Eastman Kodak Company. 1975–2011. *Annual Reports.*
  Rochester, NY: Eastman Kodak Company.
- Fforde, Gervase. 2018. *Black Wednesday and the ERM Crisis.*
  Cambridge Economic History Papers.
- Gross, Bill. 2015. "The Single Biggest Reason Why Start-
  Ups Succeed." *TED Talk,* April 2015.
- Lucas, Gavin. 2021. *Photographica: The Fascinating History of
  the Camera.* London: Bloomsbury.
- Sasson, Steven. 2013. "A Brief History of the Digital
  Camera." *IEEE Spectrum,* June 2013.

- Shakespeare, William. 2003. *Julius Caesar.* Edited by David Daniell. London: The Arden Shakespeare.
- Smith, Peter. 2014. "The Falklands War Logistics Timeline." *Journal of Military History* 78, no. 2: 589–612. https://doi.org/10.1353/jmh.2014.0096
- U.S. Securities and Exchange Commission. 1999–2011. *Kodak Form 10-K Filings.* Washington, DC: SEC.

# PRINCIPLE 17
## ADAPT QUICKLY TO CHANGING CONDITIONS

*"The reed that bends in the wind is stronger than the mighty oak which breaks in a storm."* — Aesop

The previous principles explained how to *Exploit the Environment* and *Time Your Actions Strategically*. But environments and timelines are not static. This principle asserts that in fluid, unpredictable conflicts, victory belongs to the side that adapts to changing conditions fastest. Responsiveness is not just a virtue; it is a weapon.

At the core of adaptive dominance is a simple structural truth: any adversary that is slower to change becomes vulnerable. Whether that vulnerability is exploited quickly or not, it compounds over time. The more dynamic one force grows, the more brittle its opponent becomes—until collapse occurs without a single decisive blow. Strategic superiority under conditions of flux belongs to the one who changes faster and more coherently than the other can track.

True adaptation involves more than agility. It requires infrastructure, culture, and cognition that together foster continuous responsiveness. Adaptation demands attentiveness, tolerance for ambiguity, and the courage to discard outdated but familiar tools.

The ones who survive are not those who cling to tested methods, but those who learn faster than the problem evolves.

Modern conflict dynamics make this principle non-negotiable. Climate shifts, market volatility, political realignments, evolving pathogens, generative adversarial networks, and real-time information war have collapsed the buffer time between change detection and consequence. In cyberwarfare, milliseconds matter—intrusion detection systems must flag and block malicious packets before worms spread across networks, or exploits succeed in racing system defenses. In medicine, delayed adaptation to mutating pathogens costs lives. In nature, hesitation can mean death. Across all domains, the margin for delayed responsiveness is shrinking.

Effective adaptation operates on four interlocking gears:

1. **Early Detection**
2. **Decision Speed**
3. **Resource Allocation**
4. **Iterative Learning**

Each of these mechanisms contributes to an adaptive tempo that outpaces the adversary's cycle of recognition and response.

## Early Detection: Recognizing Change Before It Breaks You

The first requirement of adaptive action is the ability to detect change before it becomes irreversible. This is not about predicting the future —it's about *perceiving* it quickly enough to shift before others do. In practice, this means cultivating sensitivity to emerging trends, faint signals, and subtle anomalies. A crack in the ceiling is not yet collapse, but it is a warning. Adaptive agents monitor the ceiling.

Detection requires systems designed for proactive scanning. In business, this could include real-time analytics, competitive intelligence, or customer behavior mapping. In the natural world, animals with acute sensory perception—like deer attuned to the snap of a distant twig—gain vital seconds to flee. In cybersecurity, threat intelligence platforms continuously parse open-source signals and dark web chatter for indicators of compromise. In all these cases, passive awareness is not enough; active environmental sensing must be built into the system's structure.

Those who fail to see early warning signs often attribute their downfall to external disruption. But disruption only breaks those who refused to bend.

### Decision Speed: A Weapon in Itself

Recognizing change is only half the battle. The window between recognition and decision is where most systems fail. Adaptive response requires judgment—often under conditions of limited information, ambiguity, and fear.

Colonel John Boyd's OODA loop (Observe, Orient, Decide, Act) remains one of the most enduring conceptual models for *Decision Speed*. It emphasizes that the advantage lies not in a perfect decision, but in a faster one. Speed disrupts. Speed forces your adversary into a reactive mode. Speed allows your system to cycle through successive adaptations before the adversary has recalibrated to the last one.

In war, the pilot who executes a maneuver faster—even if suboptimal—gains position. In litigation, the lawyer who reframes the argument first sets the discursive terrain. In politics, candidates who reorient their message fastest after a scandal often survive. In pandemics, governments that issued lockdowns, testing, and supply rerouting early—even imperfectly—fared better than those that waited for certainty.

The strength of adaptive decision-making lies not in flawless foresight but in *compounding accuracy through feedback*. A slightly wrong decision now, adjusted early, can outperform a perfect decision made too late.

## Resource Allocation: Adaptive Agility Requires Structural Flexibility

Once a decision is made, can your system move with it? The ability to rapidly shift energy, people, and tools into new configurations is what determines whether adaptation remains theoretical or becomes functional.

Rigid organizations—trapped in legacy hierarchies, approval bottlenecks, or cultural inertia—often recognize the right move but cannot execute it. The weight of their own architecture paralyzes response. In contrast, adaptive systems decentralize authority, preauthorize contingency decisions, and train for improvisation. They prepare resource modules in advance, ready to be repositioned on short notice.

In biological systems, adaptation may take the form of metabolic reprogramming. In business, it may mean shutting down underperforming products to reinforce a new direction. In AI training, it means shifting model weights or reconfiguring algorithms in response to performance drift. Without the capacity to *move*, knowledge of the right move becomes irrelevant.

## Iterative Learning: Adaptation as a Cycle, Not a Moment

Adaptation is not an event. It is a pattern. A system that adapts once may survive a crisis. A system that adapts *repeatedly and intelligently* gains cumulative advantage over time. This principle holds even in the face of failure. A flawed adaptation, if captured and reflected upon, becomes an input into future strength. But only if the system is designed to learn.

After-action reviews, version control, institutional memory systems, and feedback-driven design all support this function. They convert lived experience into structural resilience. Critically, systems must avoid scapegoating and denial during postmortems. Blame freezes learning. Adaptive organizations normalize failure as part of the learning loop—not as a crisis of identity, but as a price of agility.

## Positive Example — Kudzu's Invasive Success

Kudzu, the invasive vine that now dominates parts of the south-eastern United States, offers a clear, if unintentional, illustration of the four core mechanisms of adaptive advantage: *Early Detection* of change, *Decision Speed, Resource Allocation,* and *Iterative Learning.* While not intelligent in the strategic sense, kudzu's evolved traits allow it to behave as an extraordinarily adaptive system—outpacing and outlasting competitors not through strength or planning, but through responsiveness embedded in its biology.

**Early Detection:** Kudzu is highly sensitive to its environment. It responds quickly to gradients in light, temperature, and moisture—seeking out openings in disturbed or underutilized terrain. This environmental sensitivity allows it to expand toward opportunity faster than surrounding plants can react, exploiting newly available ecological space before others can occupy it.

**Decision Speed:** Since Kudzu is not a thinking agent, there is a caveat here. "Decisions" occur through adaptive behavior. Kudzu does not rely on a single strategy. It spreads through both seed and vegetative growth. When seed dispersal proves ineffective, it automatically leans on ground-layer propagation—rooting at nodes and forming interconnected vine networks. When cut back or burned, it regenerates from underground reserves. These shifts are not conscious decisions, but they functionally mirror adaptive responsiveness under uncertainty.

**Resource Allocation:** Kudzu rapidly redirects energy based on local context. Deep root systems extract water unavailable to competitors. Vines stretch toward vertical supports, sunlight, or open ground depending on what's most accessible. Energy is not invested evenly, but dynamically—favoring whatever path allows the plant to advance. This resource fluidity supports continuous movement into new territory.

**Iterative Learning:** Over time, certain kudzu populations have exhibited greater tolerance to herbicides or poorer soil. These traits emerge not from intentional learning but from variation and selection at scale. Kudzu reproduces quickly, and favorable traits are naturally retained. The result is a kind of structural memory—an evolving library of adaptations that compound over generations.

Kudzu doesn't plan. But it adapts—reliably, quickly, and persistently. Its structural configuration enables it to respond to change faster and more flexibly than the static systems designed to contain it.

The lesson is not to admire kudzu—it's to understand what makes it effective. Systems that are built to sense change early, adjust without central coordination, reallocate effort fluidly, and learn across cycles will outperform those that are optimized for stability. Kudzu's rise was not a fluke. It was the natural outcome of a system designed to move with its environment faster than the environment could shut it down.

## Negative Example — OpenAI's Hide-and-Seek Failure

OpenAI's 2019 multi-agent hide-and-seek experiment demonstrated what appeared to be emergent adaptation—but in reality, revealed the limits of learned behavior in a static environment. It also provides a clear illustration of failure across the four core gears of adaptive response.

The simulation involved AI agents—hiders and seekers—given no preset tactics, only rewards for outcomes. Over time, agents developed creative behaviors: hiders used movable blocks to lock themselves into rooms; seekers learned to manipulate ramps to bypass obstacles. These escalating tactics gave the impression of real-time adaptation. But when researchers introduced slight environmental changes—new tools, altered layouts, modified movement rules—the system broke.

**Early Detection:** The agents didn't register that their world had changed. Their sensors and internal models had been optimized for the original setup and could not detect novel elements outside those patterns. What looked like flexibility was actually overfitting.

**Decision Speed:** Even as their success rates dropped, agents continued attempting behaviors that no longer worked. Their policy networks had no mechanism to reassess tactics under uncertainty. The moment their assumptions were invalidated, they had no playbook for revision—only repetition.

**Resource Allocation:** Agents continued guarding objects that no longer mattered. They invested energy in outdated strategies, unable to shift effort toward exploring alternatives or evaluating new cues. Their internal prioritization remained static, even as the terrain shifted beneath them.

**Iterative Learning:** The agents could not recover from failure. There was no capacity to unlearn ineffective behavior, no memory structure for tracking new outcomes, and no system for reflecting on misalignment between action and result. The environment changed; the system degraded.

The episode revealed a critical insight: apparent adaptability isn't enough. What matters is the capacity to detect change early, revise decisions under ambiguity, shift effort dynamically, and improve across cycles. The original agents failed—not because they lacked intelligence—but because they had never been trained for variability. Later experiments showed that when exposed to randomized environments and adversarial conditions during training, agents could indeed develop more robust, transferable behaviors. In that sense, the failure wasn't final—it was diagnostic. It showed that true adaptability requires not just reaction, but systems explicitly designed for continual change.

True adaptation is not the novelty of behavior—it's the system-level capacity to shift when the rules change. These agents lacked that capacity—and that's the lesson.

### Principles in Action — Reflection Questions

1. When conditions around you shift rapidly, what signals does your system rely on to detect that change before consequences escalate?

2. In high-pressure environments, how quickly can your team or process respond with a meaningful new course of action—and who has the authority to make that call?

3. How easily can your organization redirect energy, personnel, or focus when a sudden threat or opportunity arises outside of your current plan?

4. After failed responses or surprising outcomes, what structures are in place to capture lessons and prevent repeated error?

5. What parts of your current success are built around conditions that might no longer exist tomorrow—and how exposed would you be if they disappeared?

## Sources

- Aesop. 1998. *Aesop: The Complete Fables.* Translated by Olivia and Robert Temple. London: Penguin Classics.
- Forseth, I. N., and A. F. Innis. 2004. "Kudzu (*Pueraria montana*): History, Physiology, and Ecology Combine to Make a Major Ecosystem Threat." *Critical Reviews in Plant Sciences* 23, no. 5: 401–13. https://doi.org/10.1080/07352680490505150
- OpenAI. 2019. "Emergent Tool Use from Multi-Agent Interaction." *OpenAI Blog,* September 17, 2019.
- Osinga, Frans. 2007. *Science, Strategy and War: The Strategic Theory of John Boyd.* London: Routledge.

# PRINCIPLE 18
## STRETCH THE ADVERSARY'S DEFENSES

*"He who defends everything defends nothing."* — Frederick the Great

Every defender—whether an army, an antivirus suite, or a lone parent watching a toddler (*Adversarial Dynamics in Parenting* is a book idea for another day)—has a finite stack of awareness and energy. Each new threat siphons a slice from that stack. Add enough simultaneous, credible dangers, and the defender's stack can be depleted long before the crisis ends. Frederick the Great understood this in the age of muskets. The principle never changes: with every new front an opponent is forced to defend, the vigor behind each individual defense is incrementally reduced.

This principle is the natural extension of *Targeting Weakness*. While your adversary is looking for your vulnerabilities, you can create new ones by forcing them to spread their resources too thin. However, this strategy comes with a crucial condition: it works best when you can exploit an economic mismatch in flexibility or cost. A smaller, more agile force can stretch a larger, more rigid one by forcing it to spend massively to defend multiple points. But if a small force tries to stretch a larger or equally agile one—imagine the 300 Spartans attempting to stretch the entire Persian army—that force will be annihilated. The goal is not to divide your own strength, but to force the adversary to divide theirs in a way that is disproportionately costly for them.

Stretching an adversary's defenses is not just about geography; it is a multi-layered campaign. The most effective campaigns create dilemmas by attacking across four distinct fronts, often simultaneously:

**The Spatial Front.** This is the most intuitive form of stretching. By attacking or creating threats in multiple physical or digital locations at once, you force the defender to divide their forces. Guerrillas do this by striking in separate districts; a cyberattacker does this by launching a DDoS attack against a company's public website while simultaneously attempting to breach their internal systems in different time zones.

**The Modal Front.** This involves using different categories of attack, forcing the defender to engage different types of expertise and resources. A corporation can be stretched across a modal front by facing a lawsuit (*legal mode*), a social media boycott (*PR mode*), and a targeted pricing war (*economic mode*) all at the same time. Responding requires them to coordinate their legal, communications, and sales teams, multiplying internal friction.

**The Temporal Front.** This is about manipulating time and tempo. By launching attacks in overlapping waves—one at dawn, another before the first is resolved, and a third as they regroup—you disrupt the adversary's decision-making cycle. The defender is forced to constantly re-evaluate and re-plan, burning cognitive and material resources without ever regaining the initiative.

**The Cognitive Front.** This involves exploiting the psychology of over-extension. Cognitive scientists call it *decision fatigue*: as choices mount, humans default to simpler, often incorrect, patterns. By presenting multiple, simultaneous threats—even if some are feints—you can overwhelm the defender's executive function. They may freeze, misclassify the severity of a real breach, or deploy their best assets to the wrong place.

~

WITH THESE MECHANICS LAID BARE, LET'S STEP ONTO TWO VERY different playing fields: an SEC football turf where a revolutionary offense tore an elite defense apart, and a continent where a tragically narrow defense failed to corral one toxic amphibian.

### Positive Example — Urban Meyer's Spread Offense

I'll admit it hurts my Seminole heart to praise a Florida Gator, but credit where it's due. Urban Meyer's mid-2000s Spread Offense rewrote SEC calculus by weaponizing the *Spatial Front*. Traditional power formations packed linemen into a compressed tackle box, inviting defenses to respond with equal mass. Meyer flipped that geometry. By deploying four or five wide receivers, he forced defenses to defend the entire width of the field, creating thin spots and favorable one-on-one matchups for his athletes in space.

The 2006 SEC Championship Game against Arkansas put this strategy on full display. The Razorbacks' powerful defense was stretched thin from the outset. This problem was compounded by Meyer's attack on the *Modal Front*: the triple-threat of Chris Leak's passing, Percy Harvin's speed, and freshman Tim Tebow's short-yardage power runs meant Arkansas had to defend against three different types of attack on any given play. A key second-quarter touchdown illustrated the dilemma: Chris Leak hit Percy Harvin for a 37-yard score, forcing Arkansas to honor the pass across the width of the field.

Meyer added to the pressure by attacking the *Temporal Front*. He used play tempo and hurry-up sequences that limited substitutions and complicated sideline calls. The combined effect of these three fronts was an assault on the *Cognitive Front* of the opposing coaching staff. The constant pressure—spatially, modally, and temporally—forced the defense into a state of reactive guesswork, increasing the likelihood of misalignment and fatal mistakes.

The Gators kept hammering the stretched defense, winning 38-28 and setting up their national title run. Defenses eventually adapted by recruiting hybrid players who could cover both run and pass, but that very adaptation proved Meyer's point: his multi-front attack forced opponents to change their entire philosophy. A single offense had stretched the entire conference's roster math, dictating the terms of engagement and leaving defenders sprinting from one fire to the next.

### Negative Example — Australia's Cane-Toad Containment

In 1935, Queensland's Bureau of Sugar Experiment Stations imported cane toads (*Rhinella marina*, then *Bufo marinus*) from Hawai'i to control beetles in sugarcane. An initial shipment of just over a hundred toads was bred in captivity and their progeny released widely across the cane districts. Within years, the species established and began spreading beyond cultivated fields.

For decades, official responses were fragmented and largely local: community culls, small-area barriers around sensitive sites, and opportunistic egg or tadpole removal. There was no coordinated, multi-front strategy at national scale until much later, by which time the invasion front had advanced across northern Australia.

The failure was structural across all four fronts:

**Spatial Front.** Static, locality-bound measures could not address an invasion moving across thousands of kilometers of diverse terrain wet tropics, floodplains, and savannas riddled with ephemeral waters and culverts. As range expanded, toads exploited whatever hydrological pathways were available, slipping past piecemeal defenses. Compounding the problem, the invasion *accelerated*: field studies show the front's speed increased roughly fivefold over time, driven by the evolution of longer-legged, faster-moving toads at the leading edge. In many regions the advance now occurs on the order of tens of kilometers per year.

**Modal Front.** Early efforts leaned heavily on physical removal of adults—one mode of control—while neglecting others (hydrological access, larval life-stage targeting, predator–prey dynamics). Later work demonstrated that cutting off access to *artificial* water points in arid and semi-arid landscapes can sharply reduce toad presence and native-fauna impacts—showing that a different mode (infrastructure control) can be far more leverageable than killing adults alone.

**Temporal Front.** The toad's life history outcycled management cadence. Wet-season pulses produce large cohorts that leapfrog last season's control line; without measures that persist through seasonal windows (e.g., dry-season water-point exclusion or continuous larval trapping), one-off culls simply couldn't keep pace with the invasion's rhythm.

**Cognitive Front.** Authorities treated the problem as a localized nuisance rather than a continental-scale, evolving invasion system. Only later did programs embrace research-driven tactics that exploit the toad's *own* biology—like chemical lures based on conspecific egg cues that draw tadpoles into traps—an approach that directly attacks the larval bottleneck instead of dissipating effort on adult culls.

~

THE IRONY IS THAT MULTI-FRONT INTERVENTIONS—RESTRICTING ACCESS to invasion hubs (water points) and targeting larval stages with species-specific lures—have shown strong, scalable promise where they've been deployed. But those came after the population had already surged across the Top End. Australia didn't so much "lose" to the cane toad as it tried to defend a continental problem with local, single-front tactics—until the invasion's speed, scope, and seasonality made the mismatch undeniable.

## Principles in Action — Reflection Questions

1. How could we stretch our adversary across the *Spatial Front* by threatening multiple locations, or the *Modal Front* by using different forms of attack, to force them to divide their core strength?

2. What sequence of actions on the *Temporal Front* would most effectively disrupt our opponent's decision-making, and how can we use feints or diversions to attack their *Cognitive Front* and induce decision fatigue?

3. How will we measure our opponent's response in real time? What specific indicators will tell us that our attacks on the *Spatial* or *Modal* fronts are successfully forcing them to dilute their defenses?

4. What is the cheapest, highest-leverage action we can take on any front—a social media rumor (*Cognitive*), a minor legal challenge (*Modal*), or a probe of a secondary location (*Spatial*)—that would force a disproportionately expensive defensive response from our adversary?

5. Have we identified the point at which stretching the adversary across multiple fronts begins to dangerously dilute our own forces, and what is our plan to consolidate if they refuse to be drawn out?

## Sources

- Boyd, John. 1986. *Patterns of Conflict.* Briefing slides.
- Connelly, Bill. 2019. "Blueprint for Modern Offenses." *ESPN Analytics Notebook,* August 20, 2019.
- DeVore, Jennifer L., Rebecca J. Cramp, Richard J. Capon, Michael R. Crossland, and Richard Shine. 2021. "Chemical Cues from Cane Toad Eggs Attract Conspecific Tadpoles: A Basis for Species-Specific Trapping." *Proceedings of the National Academy of Sciences* 118 (24): e2024296118. https://doi.org/10.1073/pnas.2024296118
- Florance, David, John K. Webb, Thomas Dempster, Michael R. Kearney, Amy Worthing, and Mike Letnic. 2011. "Excluding Access to Freshwater by an Invasive Amphibian Can Halt Its Spread." *Proceedings of the Royal*

*Society B*278 (1723): 3663–70. https://doi.org/10.1098/rspb.
2011.0286

• Frederick II (the Great). 1760. *Military Instructions for the Generals of His Army.*

• Lever, Christopher. 2001. *The Cane Toad: The History and Ecology of a Successful Coloniser.* London: Westbury Academic.

• Meyer, Urban, and Wayne Coffey. 2015. *Above the Line: Lessons in Leadership and Life from a Championship Program.* New York: Penguin.

• Phillips, Benjamin L., Gregory P. Brown, and Richard Shine. 2006. "Invasion and the Evolution of Speed in Toads." *Nature* 439 (7078): 803. https://doi.org/10.1038/439803a

• Shine, Richard. 2018. *Cane Toad Wars.* Berkeley: University of California Press.

• Southeastern Conference. 2006. *2006 SEC Championship Game Book.* Atlanta: Southeastern Conference, December 2006.

# PRINCIPLE 19
## DECEIVE TO FORCE MISTAKES

*"All warfare is based on deception: when able, appear unable; when active, appear inactive."* — Sun Tzu

"**A**ll" seems a pretty strong word from Sun Tzu in the opening quote. If someone has followed Principle 3 and built capability that renders strategy irrelevant, deception would be unnecessary. However, such an asymmetric power dynamic is rare in reality. Therefore, we won't disparage Sun Tzu's word choice. In matches of power parity, deception is exceedingly important.

Any adversarial contest—whether waged with battalions, court filings, or playoff rosters—begins in the mind. Because human attention and analytic energy are finite, opponents must make choices about what to watch, what to ignore, and where to spend precious resources. Deception weaponizes that scarcity. By feeding the enemy an internally consistent—but ultimately false—picture, the strategist hijacks the adversary's scarce cognition, redirects material assets to irrelevant fronts, and preserves one's own strength for a decisive stroke. The magician's empty left hand, held under the spotlight, is no different from dummy tank parks along the Kent coast—or a three-receiver formation that screams "pass" while the real angle of attack waits somewhere else.

When deception works, the payoff is asymmetric: a forged courier letter moves entire panzer divisions; a well-timed rumor crash-lands a hostile takeover bid; a telegraphed slant route gifts a Super Bowl title to the other sideline. But deception is fragile. Neglect it, and you hand adversaries unfiltered access to your crown jewels. Botch it, and you burn credibility, squander your best assets, and sometimes lose everything in a heartbeat. This principle demands mastery of both extremes: how to design a lie so elegant that opponents choose ruin themselves—and how to avoid becoming your own first victim.

## The Five-Step Craft of Deception

1. **Design.** Every ruse starts with empathy. What specific misjudgment do we want the adversary to make, and what narrative will shepherd them toward it? Without a target error—a division redeployed, a defender biting on a pump-fake, an auditor waving through the wrong column —deception becomes self-indulgent theater.

2. **Planting.** First impressions anchor belief. Therefore, the illusion must arrive via channels the adversary already trusts: a neutral-port coroner's report, a familiar financial disclosure form, a practiced short-yardage formation. If early details feel off, skepticism blooms; if they feel comfortably ordinary, the audience relaxes into complacency.

3. **Reinforcement.** Minor details matter more than grand flourishes. Ticket stubs, love letters, realistic packet timing inside a spoofed data stream—all serve to dull the analytical knife. The audience is disarmed by clutter that looks too mundane to fake.

4. **Monitoring.** A ruse without feedback is self-deception. Ultra intercepts confirmed when Germans

swallowed Mincemeat, which we'll unpack below; heat-map telemetry reveals whether a defense slides toward a decoy motion. Strategists who fail to watch the enemy's reaction inevitably out themselves.

5. **Adaptation.** Every deception decays as the adversary learns. Refresh dummy tank parks weekly; rotate malware signatures hourly; retire a football play the moment film study exposes its cues. Static deception is brittle deception.

Risk shadows every step. Legal and ethical boundaries mark hard stop lines: perfidious misuse of humanitarian symbols, fraudulent securities filings, or medical misinformation. A well-built plan includes a fail-open clause: if the mask slips, either the fallback action still yields value or the withdrawal costs are bearable. A bad plan leaves you naked.

**Positive Example — Operation Mincemeat**

For anyone versed in military history, Operation Mincemeat is no doubt *cliché*. It is one of the most-used and best-studied examples of strategic deception. But it is *cliché* for a reason: it is simply too beautiful an example not to share here.

By early 1943 the Allies had cleared North Africa and were staring at the obvious next move: Sicily. Precisely because it was obvious, Sicily bristled with guns and garrisons. British planners didn't need the Axis to be blind; they needed them to look hard in the wrong place.

*Design.* The scheme was to create a courier who never existed—"Major William Martin, Royal Marines"—and have him "carry" documents pointing to invasions of Greece and Sardinia while casting Sicily as a feint. To make the identity breathe, his pockets held ordinary clutter: an affectionate photograph of "Pam" with a handwritten note, theater ticket stubs, a sharp letter from his bank about an overdraft, and a jeweler's bill—along with letters that referenced shore leave and routine worries. That pocket litter made the extraordinary seem routine.

*Planting.* On April 30, 1943, HMS Seraph surfaced off Huelva, Spain, a coastline where German agents had access to sympathetic officials. The body, chained to a briefcase of "top secret" letters, washed ashore; Spanish authorities processed the case through channels the Abwehr already trusted.

*Reinforcement.* London behaved exactly as a government would if genuine secrets had been lost—urgent diplomatic notes pressed Madrid for the papers' return—while the broader Mediterranean deception framework (Operation Barclay) supplied consistent background noise so the story read like context, not a stunt.

*Monitoring.* British codebreakers watched for enemy reaction. Ultra decrypts soon showed Berlin treating the material as authentic: forces were postured toward Greece and Sardinia, and Sicily did not receive equivalent priority. With confirmation in hand, the team resisted the urge to embroider the tale further.

*Adaptation.* Because every ruse decays with exposure, the planners stuck to a disciplined script—no last-minute flourishes, no extra signals—letting the false picture mature without calling attention to itself.

At dawn on July 10, 1943, Allied troops landed on Sicilian beaches and met defenders strong enough to fight but too thinly spread to stop the assault. Within weeks, the campaign helped topple Mussolini (July 25, 1943); by August 17, Axis forces had evacuated the island. Mincemeat worked because each link—*Design, Planting, Reinforcement, Monitoring,* and *Adaptation*—held under strain, turning a tidy fiction into a strategic fact.

### Negative Example — Xerox PARC and Super Bowl XLIX

We must break tradition here somewhat and offer up *two* negative examples. The first represents a failure to apply Principle 19 *at all*. The second represents a failure to apply it *well*.

*Xerox PARC and Steve Jobs:*

Let's look at Xerox first. In December 1979, Apple engineers walked into Xerox's Palo Alto Research Center and were shown the future: the Alto's graphical interface, mouse-driven pointing, overlapping windows, and live, editable text in the Smalltalk environment. The visit was not espionage; it was scheduled—part of a deal granting Xerox an option to purchase Apple shares. What failed here was not invention but tradecraft.

*Design.* There was no intent to mislead a likely competitor about Xerox's crown jewels. The working assumption inside PARC and at corporate headquarters was that the gap from lab prototype to mainstream product would protect Xerox's lead.

*Planting.* The demonstrations were genuine, full-fidelity walkthroughs conducted by trusted researchers in their own labs. Apple saw real prototypes, real workflows, and the authentic speed of interaction—not a staged or throttled facsimile.

*Reinforcement.* Follow-on conversations, whiteboard explanations, and additional access reinforced the same picture: Xerox possessed a coherent, integrated GUI paradigm awaiting commercialization. Each additional bit of openness made the revelation stickier and more actionable for Apple.

*Monitoring.* Xerox did not build a live "reaction dashboard" around Apple's next moves. As Lisa (announced 1983) and Macintosh (January 1984) took shape, there was no systematic early-warning mechanism—no market sensing to adjust secrecy, no counternarrative to slow partner enthusiasm, no licensing firewall to bound what had already been shown.

*Adaptation.* Xerox's own GUI product, the Star (launched April 1981), arrived expensive and narrowly targeted, with marketing and distribution misaligned to the personal-computing wave. By the time adaptation began in earnest, Apple and then Microsoft had operationalized the very concepts PARC revealed—at scale and at consumer price points.

The lesson is not that openness is bad; it's that failing to *Design* even minimal misdirection, to curate what you *Plant,* to control the cues that *Reinforce* belief, to *Monitor* how quickly a rival is converting insight into product, and to *Adapt* your own release and go-to-market in response can turn a lab advantage into an adversary's launch pad.

*Seattle Seahawks, Super Bowl XLIX: The Slant Heard 'Round the World:*

Picture it: February 1, 2015, Glendale, Arizona. Seattle trailed New England 28–24 with 26 seconds left, second-and-goal from the 1, one timeout remaining. Marshawn Lynch had just barreled to the doorstep. Everyone in the stadium expected power run—so the Seahawks called a quick, inside pick/slant from a compact set, hoping the Patriots would sell out to stop the dive.

*Design.* The core idea—show run strength, throw where bodies are densest—can work *if* the picture forces defenders to honor the run first. Seattle's design undercut itself: the chosen personnel and compressed alignment offered less run "story" than a heavy set or crack-motion look would have.

*Planting.* Pre-snap cues planted the wrong narrative for a pass to succeed: no backfield motion to stress keys, no hard run action to freeze the second level. New England's study habits—drilling that exact goal-line concept during the week—meant the picture Seattle presented matched a known pass tendency.

*Reinforcement.* The snap reinforced pass, not run: immediate quarterback set, quick release point, and a slant into a congested lane. Nothing in cadence, formation shift, or backfield mesh added even a half beat of uncertainty that good deception layers on top of first impressions.

*Monitoring.* With the ball in flight in under a second, there was no time to confirm defender leverage or kill the call. Malcolm Butler broke downhill on the anticipated route he'd been coached to expect and arrived first.

*Adaptation.* After the interception, the postmortem focused on outcome ("Why not run?"), but the deeper issue was brittle deception. When film revealed that New England had anticipated the concept, Seattle did not pivot earlier in the sequence to a look that better sold power or offered a safe second read.

Deception is a craft of margins. At the one-yard line, Seattle needed the full chain—*Design, Planting, Reinforcement, Monitoring,* and *Adaptation*—to bend defender cognition for a heartbeat. Instead, each link nudged the defense toward the right answer, and one studied cornerback made history.

～

TOGETHER, THE PARC DEMO AND THE SEAHAWKS' GOAL-LINE CALL show the two ways deception fails: by omission and by mis-execution. Xerox never attempted deception at all—no *Design* to decide what to reveal, no selective *Planting* to gate access, and permissive *Reinforcement* that deepened Apple's insight—followed by absent *Monitoring* and sluggish *Adaptation* as rivals converted lab ideas into products. Seattle, by contrast, tried to deceive but inverted the cues: a *Design* that didn't sell run, *Planting* and *Reinforcement* that screamed pass, zero time for *Monitoring,* and no in-sequence *Adaptation* when the defense anticipated the concept. The lesson is symmetric: if you won't —or can't—engineer the full chain of *Design, Planting, Reinforcement, Monitoring,* and *Adaptation,* you are not practicing deception.

### Principles in Action — Reflection Questions

1. *Design*: What exact false picture do we want the adversary to adopt, and what single misallocation (unit movement, purchase, click path) will it trigger?
2. *Planting*: Through which trusted channels will that picture arrive—and in what sequence—and who owns each channel's timing and cover story?
3. *Reinforcement*: Which mundane details (metadata, timestamps, uniforms, stance cues) make the picture feel real, and who red-teams them for tells?
4. *Monitoring*: What hard signals (intercepts, price/traffic moves, alignment shifts) confirm uptake, and what thresholds trigger a pivot?

5. *Adaptation*: If skepticism appears, what's the prewritten branch plan (escalate/switch/abort), and when do we retire the ruse to avoid patterning ourselves?

## Sources

- Ambrose, Stephen E. 1994. *D-Day: June 6, 1944.* New York: Simon & Schuster.
- Isaacson, Walter. 2011. *Steve Jobs.* New York: Simon & Schuster.
- Macintyre, Ben. 2010. *Operation Mincemeat: How a Dead Man and a Bizarre Plan Fooled the Nazis and Assured an Allied Victory.* New York: Crown.
- National Football League. 2015. *NFL Game Book: Super Bowl XLIX.* February 1, 2015.
- Smith, Douglas K., and Robert C. Alexander. 1988. *Fumbling the Future: How Xerox Invented, Then Ignored, the First Personal Computer.* New York: William Morrow.
- Sun Tzu. 1963. *The Art of War.* Translated by Samuel B. Griffith. Oxford: Oxford University Press.
- Whaley, Barton. 2007. *Stratagem: Deception and Surprise in War.* Norwalk, CT: EastBridge.

# PRINCIPLE 20
## SACRIFICE TACTICALLY TO GAIN
## STRATEGICALLY

*"Sacrifice the small stones to capture the large ones."* – Aphorism from the game of Go

I n my analysis of adversarial dynamics, I have found that one of the most counterintuitive yet powerful concepts is the *strategic sacrifice*. The principle "Sacrifice Tactically to Gain Strategically" asserts that an entity can create a decisive, long-term advantage by intentionally and purposefully accepting a short-term loss. This is not about retreat, surrender, or simple cost-cutting. It is the deliberate act of giving up ground, resources, or position to unbalance an adversary, mislead their judgment, or create a larger, more valuable opportunity. It is the discipline of taking one step backward in order to take three steps forward.

This very principle is a manifestation of the dynamic I observed in the book's opening story. In that example, Dale Carnegie's truck salesman, when confronted by an antagonistic customer, made a tactical sacrifice. He sacrificed his ego and the immediate "win" of defending his own product. Instead of meeting force with force, he agreed with the customer and complimented the competitor, giving up the argumentative ground. This purposeful loss of his tactical position created a massive strategic gain: it disarmed the customer's hostility, built rapport, and created a new, larger opportunity to actually make the sale. Like the hapkido master who yields to an opponent's momentum to use it against them, the salesman made a small sacrifice to achieve a total victory.

If strategic sacrifice is so powerful, why is it so rarely and poorly practiced? In my experience, the primary barrier is psychological. The act of intentionally accepting a loss, however small, runs counter to our deepest instincts for self-preservation and advancement. It requires immense discipline, confidence, and a steadfast focus on the long-term objective. Leaders are often judged on short-term metrics, making it difficult to justify a tactical loss that may look like a failure to outside observers or shareholders. Furthermore, ego and the fear of appearing weak can make any concession feel like a defeat. It takes a truly sophisticated strategist to overcome these emotional and political pressures and execute a purposeful loss with clarity and resolve, distinguishing it from an outright blunder.

A true strategic sacrifice has three core components that separate it from a simple mistake. First, it must be *Purposeful*. The loss is not an accident; it is a deliberate choice made to achieve a specific, predefined strategic goal that is of higher value than what is being given up. Second, it must be *Asymmetrical*. The value of what you are giving up (the tactical cost) must be significantly lower than the value of the strategic advantage you aim to gain. You are consciously trading a pawn to win a queen. Third, it must be *Leveraged*. The act of sacrifice itself must create the new opportunity. It must be the key that unlocks the door to the larger victory, either by misleading the opponent, repositioning your own forces, or fundamentally changing the dynamics of the engagement.

One of the most common and powerful reasons to make a sacrifice is to seize the initiative (which is a major component of Principle 12). In any engagement, the entity that is reacting is at a disadvantage. By making a bold, unexpected sacrifice, you can shatter the existing tempo and force the adversary to respond to *your* move, on *your* terms. You give up something tangible and static, like material or position, in exchange for something intangible but dynamic: control over the flow of the conflict. In my analysis, this is often a brilliant trade. The adversary is so focused on consolidating their small, tactical gain that they fail to see that you have used their reaction to reposition your other forces for the decisive, strategic blow.

The most common forms of sacrifice involve tangible assets. In a military context, this could be a feigned retreat, where a commander gives up physical ground to lure an overconfident adversary into a prepared ambush. In business, it could be a "loss leader" strategy, where a company intentionally sells a popular product at a loss (a tactical sacrifice of profit margin) to draw customers into their ecosystem and capture a much larger strategic gain in long-term loyalty and market share. A tech company might designate as open-source a valuable piece of software, sacrificing direct revenue from that tool to achieve the strategic victory of establishing it as the industry standard, thereby controlling the entire ecosystem.

In my view, however, the most powerful sacrifices often involve intangible assets. This can be the sacrifice of *time*, by deliberately delaying an action to allow an adversary to overextend themselves or reveal their intentions. It can be the sacrifice of *simplicity*, by intentionally introducing complexity into a negotiation to confuse an opponent who thrives on straightforward deals. But perhaps the most difficult and potent of all is the sacrifice of *ego* or *dogma*. This is the willingness to abandon a long-held belief, admit a mistake, or give up a position of personal pride in order to achieve a collective victory. This requires a level of self-awareness and strategic discipline that few possess, but those who do can often achieve outcomes that are impossible for those shackled by their own pride.

A highly sophisticated application of this principle is to use a sacrifice as a tool to gather intelligence (Principle 4). By offering a piece of calculated "bait," you can test your adversary's mindset and intentions. Their reaction to the sacrifice provides invaluable data. Do they greedily and immediately take the bait, suggesting they are overconfident or perhaps desperate? Do they cautiously ignore it, suggesting they are paranoid or suspect a deeper trap? Do they respond in a way you didn't anticipate at all, revealing a facet of their strategy or psychology you were previously unaware of? The sacrifice becomes a question you pose to the adversary, and their response is the answer that can inform your entire subsequent strategy. It is a way of paying a small price in material for a large gain in clarity.

## Positive Example: Garry Kasparov's "Immortal" Rook Sacrifice

In chess—a purely intellectual adversarial arena—the strategic sacrifice is strategy, distilled. Garry Kasparov's win over Veselin Topalov at Wijk aan Zee in 1999 is a canonical illustration. Out of a tense middlegame, Kasparov detonated a forcing sequence that began with an intentional rook sacrifice—ceding raw material to rip open files, seize tempi (time/forced moves), and drag Black's king into a lethal crossfire. On a scoresheet, it looked insane: a five-point piece given up for a single pawn. On the board it reweighted the entire position in his favor.

The choice was *Purposeful*. Kasparov wasn't improvising for drama; the sacrifice targeted a single strategic objective—shatter Black's defensive coordination and expose the king. Every move that followed—the checks, the quiet intermezzos, the precise deflections—served that end. The moment the rook came off, Topalov's forces were compelled into awkward squares, lines toward his monarch pried open, and coordination—more valuable than any single unit—began to collapse.

It was also *Asymmetrical*. Measured materially, White fell behind. Measured dynamically, the ledger flipped. Kasparov converted a short-term, countable loss into long-term, compounding assets: initiative, time, and irrevocable weaknesses around the king. Those intangible advantages—initiative and king safety—scale multiplicatively. Each tempo Kasparov gained intensified threats, forced further concessions, and widened the gap between nominal material and real control of the game.

Crucially, the sacrifice was *Leveraged*. It did not merely precede the winning attack; it created it. By forcing Topalov to accept the exchange and recapture on Kasparov's terms, the position transformed: diagonals burst open for bishops, the queen infiltrated with tempo, and Black's pieces were tied to hopeless defense. From there, Kasparov's play looked effortless only because the sacrifice had engineered inevitability. Topalov eventually resigned with his king exposed and his army paralyzed—a textbook case of trading a pawn's worth of wood for a queen's worth of position. The short-term deficit bought an unassailable strategic advantage.

### Negative Example: Amy Winehouse and the Failure to Sacrifice for Sobriety

We haven't talked much about intrapersonal conflict as an adversarial domain, but I studied it extensively in my original UTAD research. And this is an appropriate, if tragic, opportunity to bring it up. Of course, addiction is a complex and often tragic condition—but within that complexity there are lessons to be learned. We analyze the situation here from a place of human love, understanding, and respect.

Intrapersonal conflict can be just as adversarial as any market or battlefield. Amy Winehouse's story pits two forces within one person: the drive to live, create, and keep making music, and the adaptive internal adversary of addiction. Friends, family, and clinicians repeatedly urged a path that required hard, near-term losses—sustained inpatient treatment, a redesigned social environment, and distance from a codependent relationship that enabled substance use. Those were the tactical positions that had to be relinquished for the larger war to be won.

The needed choices were *Purposeful*. Submitting to structured, longer-term care and stepping away from enabling ties would have been deliberate losses of freedom, momentum, and image in service of a clear strategic objective: durable sobriety and the life it protects. They were *Asymmetrical*. The immediate costs—privacy, touring cadence, short-run earnings—were small next to the strategic value of health, safety, and the ability to create over decades rather than months. And they would have been *Leveraged*. Treatment and environment change are not symbolic gestures; they alter the operating terrain. Clinical scaffolding, safer routines, and new accountability can disable the very reinforcers that keep addiction entrenched, creating options (time to heal, space to write, repaired relationships) that simply do not exist without the sacrifice.

Instead, the necessary losses went unmade. Independence—as framed by the addiction—was preserved; the social orbit and relationship dynamics that rewarded use persisted. Without the *Leveraged* shift in environment, the internal adversary retained interior lines of supply. The strategic outcome followed from those choices: despite periods of improvement, the pattern reasserted itself, and her death from alcohol poisoning in 2011 became the tragic culmination of a conflict in which refusing small, painful sacrifices foreclosed the only path to a larger victory.

## Principles in Action — Reflection Questions

1. What is the one thing—a resource, a position, a belief, or your own ego—that you are currently unwilling to give up, and could sacrificing it intentionally create a much larger strategic opportunity?
2. In a negotiation or conflict, what is the "easy win" that is right in front of you? What greater victory might be possible if you were willing to sacrifice that easy win as a calculated risk?
3. What is the tactical "bait" you could offer an adversary? What small resource or apparent vulnerability could you present that would lure them into a larger, preplanned trap?
4. Is our team or organization clinging to a profitable but obsolete product line or strategy (like Kodak's film from a prior chapter)? What would a purposeful, tactical sacrifice of that legacy asset look like, and what new strategic ground would it open up?

5. How can we distinguish between a disciplined, strategic sacrifice and a simple, panicked retreat? Does our proposed "purposeful loss" have a clear, direct, and probable path to a greater strategic gain?

## Sources

- Burgess, Graham, John Nunn, and John Emms. 2010. *The Mammoth Book of the World's Greatest Chess Games.* London: Robinson.
- Carnegie, Dale. 1936. *How to Win Friends and Influence People.* New York: Simon & Schuster.
- Kapadia, Asif, dir. 2015. *Amy.* London: On the Corner Films / Film4.
- Kasparov, Garry. 2017. *Deep Thinking: Where Machine Intelligence Ends and Human Creativity Begins.* New York: PublicAffairs.
- Newkey-Burden, Chas. 2008. *Amy Winehouse: The Biography.* London: John Blake.
- Van der Kolk, Bessel A. 2014. *The Body Keeps the Score: Brain, Mind, and Body in the Healing of Trauma.* New York: Penguin Books.

# PRINCIPLE 21
## DRAIN THE ADVERSARY'S RESOURCES
## SYSTEMATICALLY

*"We are continuing this policy in bleeding America to the point of bankruptcy."* — Usama bin Ladin

I hate giving credit to Usama bin Ladin *almost* as much as giving credit to Urban Meyer. But hear me out. Back in 2004, as a rookie counterterrorism analyst at U.S. Central Command, I felt confident we would quickly throttle al-Qa'ida with superior funding, satellites, and precision weapons. Then a mentor slid a dog-eared copy of *Imperial Hubris* across my desk. Author Michael Scheuer—who'd hunted bin Ladin from inside the CIA—argued that the terrorist leader wasn't a cave-dwelling madman at all, but a shrewd strategist intent on bankrupting the United States by provoking expensive responses to cheap attacks.

I initially scoffed—but then kept returning to Scheuer's assessment over the next eighteen years as the war costs multiplied. Between 2001 and 2022, the U.S. spent an estimated $8 trillion on the Global War on Terror. Al-Qa'ida's annual operating budget never cracked $50 million. That's an annualized resource exchange ratio of roughly 8,000:1. Scheuer's thesis felt less heresy and more elementary arithmetic: if each enemy dollar can force us to spend thousands, time itself becomes the adversary's ally.

The principle is older than terrorism. Attritional geniuses from Võ Nguyên Giap to Ulysses S. Grant understood that *cost velocity*—how rapidly one side burns resources relative to the other—can determine outcomes independent of battlefield scorecards. When you can't out-gun or out-maneuver the adversary, out-economics them. Drain their wallets, their attention span, their morale, and watch the house of cards sag under its own debt.

### The Cost-Imposition Curve

$$R = Cost\ forced\ on\ adversary \div Cost\ spent\ by\ us$$

A campaign is resource-positive when $R > 1$. Bin Ladin's IEDs: $R$ sometimes exceeded 10,000. Cyber extortionists paying $40 for ransomware kits but extracting $4 million from a hospital achieve $R \approx 100,000$. A $200 legal filing can force six figures in compliance reviews. To drive $R$ upward, you must combine asymmetric spending (cheap tools) with asymmetric *response triggers* (the enemy responds with expensive protocols).

### The Tri-Resource Drain Cycle

1. **Material Drain** – money, munitions, compute cycles.
2. **Cognitive Drain** – leadership bandwidth, analyst focus, staff burnout.
3. **Emotional Drain** – public patience, shareholder confidence, troop morale.

Material losses can sometimes be replenished if the other two tanks stay full. But when cognitive capacity erodes under 18-hour shifts, or emotional support collapses amid casualty counts, ample material will not sustain a force.

## The Tempo Sweet Spot

Attrition is not stasis; it's tempo management. Hit too infrequently and the defender replenishes between blows. Hit too fast and you expend faster than they do. The sweet spot is the defender's "recovery half-life"—the time it takes the defender to restore cash, attention, and willpower. Craft your attack cadence just inside that window, keeping them forever parched without draining yourself.

## Micro-Cost, Macro-Drain: The 5.56 mm Paradigm

The U.S. shift from 7.62 mm (M14) to 5.56 mm (M16) in Vietnam reflected small-caliber, high-velocity logic: controllable automatic fire, lighter cartridges, and far more rounds per resupply ton. In an attritional frame, that logistics math matters—more carried rounds can sustain more contact. Not only that, but a related battlefield truth amplified the effect: a wounded combatant can require more down-stream resources than a KIA (killed in action)—casualty evacuation, medics, beds, rehabilitation, replacement training. This wasn't the sole driver of the caliber transition, but it neatly illustrates cost-imposition dynamics.

## Counterattrition Safeguards

A drain strategy backfires if the attacker bleeds faster than the target. Successful practitioners:

- **Diversify funding.** Blend local levies, diaspora/online fundraising, and institutional patrons (public or private) so one disrupted stream doesn't end the campaign.
- **Automate repeats.** Use templates, scheduling, and modular tooling so recurrent tasks (filings, outreach, monitoring) scale without linear headcount.
- **Exploit frugality.** Favor robust, repairable gear and lightweight logistics that survive dirt, distance, and delay;

avoid brittle high-spec inputs that create spare-parts debt.

Defenders break drain spirals by abstraction and substitution: replace man-hours with autonomous sensing, swap overtime triage for AI-assisted routing, and crowdsource expertise (e.g., pro bono legal clinics) to flatten the attacker's $R$ ratio.

## Positive Example — How the Soviet-Afghan War Became a Decade-Long Hemorrhage

On December 24, 1979, Soviet forces began large-scale airlifts and border crossings into Afghanistan to stabilize a faltering client regime; three days later, Storm-333 seized the presidential palace and killed Hafizullah Amin. What Moscow expected to be a short stabilization evolved into a grinding counterinsurgency that bled money, leadership bandwidth, and public will.

*Material Drain.* Sustainment defined the war. The 40th Army required constant convoy escorts for fuel, ammo, and spares over mountain roads vulnerable to mines and ambush. Rotary-wing support—the Mi-24 "Hind" fleet—consumed vast aviation fuel and maintenance hours. When FIM-92 Stinger man-portable air-defense systems (MANPADS) arrived in late 1986, the cost-imposition ratio spiked: missiles costing tens of thousands could down airframes worth millions or force operational changes (higher altitudes, more flares, more sorties for the same effect). Even when aircraft weren't shot down, the protective adaptations (altitude, routing, countermeasures) raised per-mission cost and lowered effectiveness—classic $R > 1$ in the defender's favor.

*Cognitive Drain.* Rotations churned institutional memory. Commanders cycled roughly annually, relearning terrain, local power networks, and IED patterns from scratch. Staff attention tunneled into convoy protection, mine-clearing SOPs, and village cordon routines—leadership bandwidth that couldn't be spent on war termination or political strategy. The enemy's playbook stayed simple; the occupier's got more complex every quarter.

*Emotional Drain.* Over the decade, about 620,000 Soviets served; roughly 15,000 returned in coffins and 50,000+ were wounded. Televised funerals, maimed veterans, and a widening credibility gap at home eroded social license for an open-ended war. By the late 1980s, senior leadership, including Gorbachev—who called Afghanistan "the bleeding wound"—concluded the political and moral accounts were overdrawn, independent of tactical reports from the field.

The ledger at exit captures systematic drain: the Soviets withdrew in 1989 after expending tens of billions of dollars and irreplaceable human capital, while the insurgency—fueled by comparatively modest external funding and local support—had repeatedly forced an $R$ far above 1 through cheap attacks that compelled expensive Soviet responses.

## Negative Example — How Amazon Leveraged Deep Pockets to Exhaust Quidsi (Diapers.com)

Quidsi (founded 2005) built a profitable niche shipping bulky, low-margin diapers with smart logistics and sticky customers. By 2009, revenue had climbed into the hundreds of millions, and the category looked strategically important to Amazon.

*Material Drain.* Amazon unleashed a price-matching algorithm— "match or beat Diapers.com on every SKU"—and layered Amazon Mom (Subscribe & Save plus steep promotional discounts). Because Amazon's buyer power and fulfillment efficiency drove its unit costs below Quidsi's, the same sticker-price cut typically cost Quidsi more to match than it cost Amazon to make—$R > 1$ by design. Add heavy Google AdWords spend, and Quidsi's customer-acquisition cost jumped while gross margin compressed toward zero. Amazon could absorb nine-figure losses in a single vertical; Quidsi could not, and its burn accelerated to roughly $10 million per month.

*Cognitive Drain.* Quidsi fought with 24-hour repricing sprints and manual catalog tweaks—humans sleeping in conference rooms to keep up—while Amazon's automation did the bulk of the work. Leadership time bled into firefighting: which SKUs to match, which to abandon, which promos to honor. Every hour spent on tactical price defense was an hour not spent on new categories, partnerships, or financing strategy—another channel where Amazon's dollar of automation imposed more than a dollar of opponent effort ($R > 1$ in attention).

*Emotional Drain.* As the burn rose and the narrative hardened—Amazon would keep prices uneconomic until a sale—investors' confidence wavered, lenders balked, and employee morale sagged. Capital got scarcer and costlier for Quidsi even as Amazon's war chest made its own spend feel routine; one side's anxiety became the other side's leverage.

Outcome: In November 2010, Amazon acquired Quidsi for roughly $545 million. The brands continued for several years, but by 2017 Amazon shut down remaining operations and folded the know-how into its stack. The lesson isn't just that a giant can spend more; it's that material burn, cognitive overload, and emotional erosion can be orchestrated so that not selling becomes the least rational option for a smaller rival.

## Principles in Action — Reflection Questions

1. Where can $1 of our spend force at least $10 of theirs—and how will we verify that delta in cash, attention, or sentiment?
2. What is the defender's recovery half-life, and what attack tempo keeps them below it without steepening our own cost curve?
3. Which repeatable tasks can we automate or templatize so our marginal cost trends toward zero as volume rises?
4. What leading indicators will tell us which tank is draining fastest—material, cognitive, or emotional—and when to pivot fronts?

5. What countermeasures or narrative backlash could flip our R below 1, and what substitutions or inoculations do we have ready to neutralize that?

## Sources

- bin Ladin, Usama. 2004. "Full Transcript of bin Laden's Speech." *The Guardian,* October 29, 2004.
- Boyd, John. 1986. *Patterns of Conflict.* Briefing slides.
- Central Intelligence Agency. 1999. "Soviet Aircraft Losses in Afghanistan, 1984–1988." Declassified memo.
- Crawford, Neta C., and Catherine Lutz. 2021. *The U.S. Budgetary Costs of the Post-9/11 Wars, 2001–2022.* Providence, RI: Brown University, Watson Institute, Costs of War Project.
- Grant, Ulysses S. 1885. *Personal Memoirs of U.S. Grant.* New York: Charles L. Webster & Co.
- Grau, Lester W., and Michael A. Gress. 2002. *The Soviet–Afghan War: How a Superpower Fought and Lost.* Lawrence, KS: University Press of Kansas.
- Manjoo, Farhad. 2010. "Amazon's Diaper War." *Slate,* November 4, 2010.
- National Commission on Terrorist Attacks Upon the United States. 2004. *The 9/11 Commission Report.* Washington, DC: U.S. Government Printing Office.
- Rasanayagam, Angelo. 2005. *Afghanistan: A Modern History.* London: I.B. Tauris.
- Scheuer, Michael. 2004. *Imperial Hubris: Why the West Is Losing the War on Terror.* Washington, DC: Potomac Books.
- Stone, Brad. 2013. *The Everything Store: Jeff Bezos and the Age of Amazon.* New York: Little, Brown.
- U.S. Army Ordnance Branch. 1962. *Small Arms Ammunition Study: Caliber Transition Analysis.* Aberdeen Proving Ground, MD: U.S. Army Ordnance Branch.

# PRINCIPLE 22
## DISRUPT THE ADVERSARY'S CONTROL CENTER

*"Strike the shepherd, and the sheep will be scattered."* — Zechariah 13:7

I f you've seen the 1996 movie *Independence Day*, then you no doubt remember the turning point in humanity's fight against the technologically superior aliens. The insight came when Jeff Goldblum's embattled genius character, David Levinson, had an epiphany: instead of fighting a hopeless battle against thousands of individual ships, they could neutralize the entire fleet by targeting its central command system. His plan was to upload a computer virus to the alien mothership, which would disrupt the control center and create chaos among all the smaller ships below.

That one idea reframes a hopeless, planet-wide slugfest into a scalpel job: break the *brain*, and every tentacled gun platform becomes scrap metal. We laugh at the popcorn physics, but the strategic spine is steel-hard. Whenever an organism—ant nest, cartel, Fortune 100 conglomerate, or alien space fleet—funnels decisions through a single nerve bundle, that bundle is the cheapest, fastest lever for decisive change. Clausewitz urged strikes against an enemy's *center of gravity*—the source of cohesion whose collapse unravels the whole cloth. This would certainly include something we may call a *control center.* Strictly, a center of gravity need not be a single node; it can be a distributed source of cohesion. But when that cohesion is mediated through a specific command or coordination hub, hitting that hub approximates a CoG strike at discount.

A control center can wear many faces:

- a queen's pheromone factory deep in an ant gallery,
- a cloud dashboard allocating rideshare drivers by surge heat map,
- a cartel accountant's Excel sheet of railcar IDs,
- an AI-driven logistics kernel scheduling drone swarms millisecond by millisecond.

Whatever its costume, the node serves three irreplaceable functions. *Unidirectional authority* pushes guidance outward; *high-bandwidth convergence* sucks raw data inward; *resource scheduling* adjudicates where labor, ammunition, or cash should flow next. Shut down that trifunction conduit and three shock waves roll through the victim network.

*Paralysis* arrives first. Peripheral agents—workers, bots, patrol lieutenants—cling to yesterday's orders or freeze in safe mode. *Fragmentation* follows: local actors riff their own agendas, spawning mutually hostile splinters. Finally comes *exploitation*: a patient attacker steps into the vacuum, rewriting loyalties or collecting undefended assets like fruit off the ground.

The ratio of cost to payoff is grotesquely favorable *if* you strike the right node, at the right instant, and exploit the crack before it heals. Miss any element and the discount evaporates. Worse, a botched decapitation gifts your adversary (1) a free demonstration of your tools, and (2) a permission slip to retaliate under the banner of righteous defense.

Control centers do not advertise themselves. Visible bosses may be figureheads; the real nerve may be a drab back office where two clerks balance ledgers. In cyber campaigns, attackers sometimes burn a zero-day on a web server only to learn the production database lives inside an air-gapped vault. Guerrilla advisers teach: *"Kill the radioman, not the shouter with the megaphone."* Mapping demands patient surveillance—subpoena chains that follow ACH transfers, machine-learning models that spot anomalous packet timing inside encrypted streams.

The moment a queen ant bleeds alarm pheromone, every exit tunnel becomes a spiked choke point. Likewise, a cartel data clerk who hears sirens can encrypt or shred drives in sixty seconds. The strike must be surgical: surprise, overwhelm, and silence in one breath. That often means *deepfakes of normalcy*—a forged API token that reads like yesterday's, a forged pheromone that smells exactly like home.

Chaos is a wasting asset. Chimps often capitalize on weakened neighbors soon after a successful raid; world-scale botnets can redeploy command-and-control (C2) addresses minutes after defenders sever one pipe. The attacker must pre-stage new commands, reserve political talking points, or preposition conventional forces, ready to pour through the gap before the defender knits a backup brain.

Put together, these habits—mapping, surgical entry, lightning follow-through—form a grim calculus: *one* well-timed puncture of the mind can spare years of costly limb-by-limb attrition. Evolution figured this out long before humans minted acronyms. The next two stories show it in action—first in the wild where nature perfects ruthless efficiency, then in a city where humans mistook noise for signal and paid in blood.

### Positive Example — Slave-Maker Ants and Chimpanzee Border Raids

I relish any opportunity to bring our domain examples back to nature. There is something pure and primal about seeing adversarial dynamics play out on this stage. We'll look at a couple of examples here.

*Polyergus* slave-maker ants march like commandos in miniature. Each summer, scouts locate a neighboring *Formica* nest, then raiders surge in to kill or displace the host queen and seize brood. They don't linger over workers or food stores; the goal is control of the reproductive "brain." Back home, the stolen pupae eclose (emerge) and— imprinted by nest odor—serve the conquerors. Decapitate the queen and the colony's cohesion dissolves; swap whom the brood serves and the entire economy changes ownership overnight.

Half a world away, in Uganda's Ngogo Forest, primatologists documented a subtler but equally chilling program: years of lethal chimpanzee raids that picked off males from a neighboring community. Small parties crossed borders, isolated single defenders, and killed quickly—then waited. As patrol density fell, the Ngogo chimps pushed farther. Over time, the rival group's cohesion fractured; females defected for protection, bringing reproductive capital, and the aggressor chimps annexed several square kilometers of fruit-rich territory. Calories outlaid: minimal. Territory and mates gained: generational.

From insects to primates, the blueprint repeats: identify the irreplaceable command node, strike it cleanly, exploit the vacuum before a successor stabilizes. Energy ROI skyrockets; collateral bloodshed plummets.

## Negative Example — The Early War on the Medellín Cartel

In the 1980s, the United States faced a cocaine epidemic fueled by the industrial-scale trafficking of Pablo Escobar's Medellín Cartel. The initial U.S. strategy was a textbook example of attacking the periphery while leaving the control center untouched. Federal and local agencies focused their efforts on the most visible parts of the network: arresting street-level dealers in American cities and interdicting cocaine shipments in the Caribbean.

DEA and Coast Guard agents posed triumphant beside tons of seized cocaine, and prosecutors announced record numbers of arrests. News reports celebrated these tactical victories. But the network's brain, Pablo Escobar and his lieutenants in Colombia, remained largely insulated from the pressure. For every boat seized, many more were sent. For every dealer arrested in Miami, a dozen more were recruited. The cartel's control center could absorb these losses indefinitely, treating them as a simple cost of doing business. The strategy of amputating the limbs was having no effect on the operational nerve center.

Evidence of the failure was stark: despite years of seizures and arrests, the price of cocaine on American streets actually fell, while its purity increased—a clear sign that the supply chain was not meaningfully disrupted. The control center was simply too efficient at replacing its losses.

Only in the late 1980s and early 1990s did strategy pivot to a direct decapitation effort—the "kingpin strategy"—that targeted Escobar himself. When Escobar was killed in 1993, the highly centralized Medellín Cartel shattered within months: the limbs had always been replaceable; the brain was not.

### Principles in Action — Reflection Questions

1. Which data streams—financial ledgers, pheromone trails, encrypted chats—prove a node actually *directs* operations rather than merely decorates them?
2. If our decapitation attempt fails, what capabilities have we just exposed to the opponent, and how quickly can they pivot their defenses?
3. How many successor nodes have we identified, and can we strike them within the same operational tempo before chaos congeals?
4. What exploitation force—narrative, logistics, troops— stands on deck to capitalize the instant the brain goes dark?
5. Could severing the node trigger humanitarian or collateral crises that rebound politically, and have we staged mitigations?

## Sources

- Bowden, Mark. 2001. *Killing Pablo: The Hunt for the World's Greatest Outlaw.* New York: Atlantic Monthly Press.
- Clausewitz, Carl von. 1976. *On War.* Edited and translated by Michael Howard and Peter Paret. Princeton, NJ: Princeton University Press.
- Hölldobler, Bert, and Edward O. Wilson. 1990. *The Ants.* Cambridge, MA: Harvard University Press.
- *Independence Day.* 1996. Directed by Roland Emmerich. Beverly Hills, CA: 20th Century Fox.
- Mitani, John C., David P. Watts, and Sylvia J. Amsler. 2010. "Lethal Intergroup Aggression Leads to Territorial Expansion in Wild Chimpanzees." *Current Biology* 20: R507–R508. https://doi.org/10.1016/j.cub.2010.04.021
- U.S. Drug Enforcement Administration. 2003. *DEA History Book: A Tradition of Excellence 1973–2003.* Washington, DC: U.S. Drug Enforcement Administration.

# PRINCIPLE 23
## BREAK THE ADVERSARY'S WILL TO FIGHT

*"It is not the towering sail, but the unseen wind that moves the ship."* —
Proverb (Variously Attributed)

Some victories are won on the battlefield. The most decisive ones are won in the mind. To break an adversary's will to fight is to dissolve the psychological foundation of resistance so that physical confrontation becomes unnecessary—or merely incidental. This principle is not about eliminating capacity; it is about collapsing intent.

The will to fight is targetable. It is not mist and mood; it is a *structure* with load-bearing elements that can be stressed, cracked, and made to fail. When it goes, armies stop marching, coalitions splinter, and movements simply ebb away. This is the most efficient victory of all: the adversary lays down arms not because they have nothing left to fight with, but because they see nothing left to fight for.

### The Three Pillars of Will

Think of will as fuel, and of *morale* as the gauge that tells you how much is left. Beneath that gauge sit three pillars that actually hold the weight:

1. **Belief in Victory** — the conviction that continued struggle can still succeed.
2. **Cohesion** — the trust and identity that bind individuals into a coordinated whole.
3. **Legitimacy** — the moral and political warrant that makes sacrifice feel meaningful to fighters and to those who enable them.

When these pillars hold, systems endure astonishing punishment. Crack one and the others begin to sag; morale is the first needle to drop.

## Diagnosing the Nature of Will

Before you pick a strategy, ask what kind of will you are facing.

- **Brittle Will** is opportunistic and transactional—held together by short-term gain or weak leadership consensus. It often collapses when costs rise.
- **Resilient Will** is rooted in sacred values or existential identity—ideology, religion, nation, or survival. Pressure can harden it,

Treating a resilient will like a brittle one invites disaster. The only thing worse than failing to break an opponent's will is reinforcing it.

## How Will Is Broken

- **Breaking *Belief in Victory*** (Inducing Hopelessness). Show that success is structurally impossible, not merely unlikely. Pair physical pressure with narrative so each failed offensive, interdicted shipment, missed quarter, or denied motion reads like tomorrow's forecast, not yesterday's fluke. Two reliable levers sit here: Attrition as Psychological Warfare (visible, accumulating losses that prove a trend) and Inevitability (calm, relentless steadiness that makes resistance feel pointless).

- **Breaking *Cohesion*** (Sowing Internal Division). Cohesion is built on aligned goals and trust under pressure; fracture it by inserting doubt and risk inside the group. Use selective outreach to sub-factions, incentives that peel moderates from hard-liners, and Deception (Principle 19) to exploit weak bonds so members fear each other more than they fear you.

- **Breaking *Legitimacy*** (Eroding Legitimacy). Legitimacy binds sacrifice to purpose. Undermine it with precise, truthful exposure of hypocrisy or harm, timed to audiences whose consent enables the fight (insiders and external stakeholders). Done well, leaders hesitate, supporters pull permission and resources; done sloppily, it backfires and unifies the other side.

### The Risk of Backlash

This blade cuts both ways. A failed bid to break will can forge a stronger will. If your pressure looks cruel or illegitimate, neutrals harden into partisans, and families of the punished become recruiters for the cause. Deep cultural and ideological understanding is not accessory work here—it *is* the work.

### Positive Example — The Otpor! Movement in Serbia

By 1998 in Belgrade, a handful of students with a spray-paint stencil and a clenched-fist logo decided to fight a dictator with jokes, training, and discipline. They called themselves Otpor!—"Resistance!"— and their antics looked unserious until they started to work. The fist appeared on walls, flyers, buttons; it felt everywhere at once. Street actions were cheeky: mock award ceremonies for "the most corrupt official," cardboard TVs with the regime's slogans flipped upside down, a barrel labeled "donations for the dictator" that passersby beat with sticks. Each gag carried a straight-faced purpose: normalize dissent, recruit, and survive the crackdown that always came next.

*Breaking Belief in Victory.* Instead of promising a single cathartic day, Otpor! built an escalator of small wins—stickering drives, teach-ins, boycotts, parallel vote tabulation—so new recruits could feel progress now, not someday. Each arrest produced a larger rally; each confiscated banner spawned a dozen more. Independent outlets amplified every incident, and precinct-level monitoring made it plain that any "victory" manufactured at the ballot box would be documented, exposed, and contested. The internal forecast inside ministries shifted from "we can ride this out" to "we cannot win this way," collapsing the regime's belief that continued repression could still secure an acceptable outcome.

*Breaking Cohesion.* Otpor! trained volunteers to remain strictly nonviolent on camera—no projectiles, no masks, no excuses—so every baton swing read as unprovoked. Activists handed roses to riot police, delivered sandwiches, and quietly leafleted officers' families with appeals to conscience and safety. Strikes and slowdowns by workers, including the coal miners at Kolubara, widened seams between hard-liners and those worried about blackouts and pay. Municipal chiefs balked at orders from Belgrade; conscripts hesitated when the faces across the line were classmates. Coordination began to look riskier than restraint, and the regime's internal trust frayed at the joints that mattered most.

*Breaking Legitimacy.* Satire turned Slobodan Milošević from feared patriarch into a ridiculous figure, and the chant *Gotov je!*—He's finished!—made the end feel overdue rather than radical. When the September 2000 vote was cooked, Otpor!'s network produced granular evidence rather than slogans. Every teenager frog-marched into a van for wearing a fist T-shirt became another affidavit in the court of public opinion. Donors, media, and clergy who had hedged began to peel away. The state still held truncheons and TV studios, but it no longer held permission.

On October 5, 2000, buses rolled into Belgrade from factory towns, a bulldozer nosed through police lines as if pushing debris, and crowds seized parliament and the state broadcaster. Crucially, most police refused to fire; the army stood aside. Capacity remained, but will did not. By the time buildings changed hands, the state's pillars—belief in victory, cohesion, and legitimacy—had already failed.

### Negative Example — The British Empire's Failure During the Salt March

Under British rule, salt in India was a government monopoly. The Salt Act barred Indians from producing or selling their own salt and forced them to buy taxed, "official" salt. In a hot climate where laborers, farmers, and families needed salt every day—to replace sweat, preserve food, and keep livestock healthy—the levy landed like a head tax on the poor. Worse, the law criminalized acts that felt natural and local: boiling seawater in a pan, scraping crystals from a dry lakebed, or simply picking up salt that the sun had left on the shore. That daily intrusion made the monopoly a perfect symbol of rule without consent.

Gandhi chose this pressure point on purpose. After warning the viceroy in early March 1930 that he would begin *satyagraha*—a civil protest—if the salt tax stayed, he picked a tactic any villager could copy at almost no cost and with almost no training: break the salt law in daylight. The act would be both literal and emblematic—collecting salt to show that an unjust law could not command obedience when it touched every kitchen in the country.

On March 12, 1930, Gandhi left Sabarmati Ashram with seventy-eight volunteers. Twenty-four days and roughly 240 miles later, at Dandi, he stooped and lifted a crust of salt from the tide line. That was the point of the gesture: if the law is unjust, break it where all can see—and make it simple enough that millions can do the same.

*Breaking Belief in Victory (Backfired).* The Raj tried to project inevitability through arrests, bans, and censorship, assuming fear would halt replication. The opposite occurred. Each detention sparked new marches; each raid on village brine pots produced more brine pots. Dispatches from international correspondents shifted the perception ledger: every *lathi*—a type of police baton—charge looked like proof the salt law could not be defended in the open, and every village that kept marching looked like a vote for inevitability. Instead of learning "we cannot win," Indians learned "we cannot be stopped," and belief in victory hardened.

*Breaking Cohesion (Failed).* Officials expected class, caste, and regional fissures to limit solidarity. Nonviolent discipline and shared symbolism did the opposite. Shopkeepers shuttered, students left classrooms, dockworkers and mill hands joined lawyers in the street; the ethic of *satyagraha* made it easier for communities that disagreed on much to agree on this. Even within the imperial apparatus, strain showed: Indian officers ordered to club unarmed neighbors hesitated, and administrators argued over how many jail cells a salt crystal was worth. Repression fused the movement's coordination and introduced doubt inside the state's chain of command.

*Breaking Legitimacy (Boomeranged).* The government framed its actions as routine law enforcement against agitators. Then came the incident in Dharasana. In May, waves of unarmed volunteers advanced on the Salt Works and were beaten to the ground in full view of reporters. The images traveled, and the moral math inverted: a power claiming a civilizing mandate was blooding citizens for evaporating seawater. Moderates in Britain squirmed; sympathizers abroad multiplied; fence-sitters at home reclassified the salt tax from obscure revenue device to referendum on the right to govern. The lever meant to deter instead delegitimized.

Britain did not lose the capacity to rule India in 1930; it lost something harder to rebuild. By misreading a resilient will as brittle and applying pressure calibrated for deterrence rather than legitimacy, the Raj pumped energy into the very pillars it needed to collapse. The road to independence was long, but the psychological direction of travel flipped that spring: an empire that held the guns began to lose the governed.

## Principles in Action — Reflection Questions

1. What is the psychological center of gravity in our adversary's system—and which pillar (Belief in Victory, Cohesion, Legitimacy) should we stress first?
2. What evidence tells us we're facing brittle will versus resilient will, and how does that change the mechanism we choose?
3. Where can steady, visible inevitability do more to break belief than sporadic high-cost punches?
4. Which sub-factions, relationships, or incentives can we safely and truthfully leverage to increase internal friction without unifying the other side?
5. What indicators—defections, messaging shifts, turnout, donor behavior—will confirm pillar damage, and what's our contingency if pressure backfires?

## Sources

- Brown, Judith M. 1977. *Gandhi and Civil Disobedience: The Mahatma in Indian Politics, 1928–1934.* Cambridge: Cambridge University Press.
- Grossman, Dave. 1996. "Defeating the Enemy's Will: The Psychological Foundations of Maneuver Warfare." *Marine Corps Gazette.*
- Liddell Hart, B. H. 1967. *Strategy.* 2nd ed. New York: Praeger.
- Popovic, Srdja. 2015. *Blueprint for Revolution: How to Use Rice Pudding, Lego Men, and Other Nonviolent Techniques to Galvanize Communities, Overthrow Dictators, or Simply Change the World.* New York: Spiegel & Grau.

- Sharp, Gene. 1973. *The Politics of Nonviolent Action*. Boston: Porter Sargent.
- Weber, Thomas. 1997. *On the Salt March: The Historiography of Gandhi's March to Dandi*. New Delhi: Oxford University Press.
- *Bringing Down a Dictator*. 2002. Directed by Steve York. Washington, DC: York Zimmerman Inc. (PBS broadcast).

# PRINCIPLE 24

## CALCULATE RISK AND REWARD IN EVERY ACTION

*"Take calculated risks. That is quite different from being rash."* — General George S. Patton

Good decisions do not come from intelligence alone. By *intelligence* here, I mean the general ability to acquire and apply ideas and skills, not the process of systematically collecting and analyzing information on an adversary. Decisions come from a clear, repeatable method for comparing likely gains with likely losses when time is short. As far as intelligence's role, I define what's needed—*intelligence that matters*—as the ability to trace cause and effect quickly and keep those links valid as complexity grows. The principle here—*Calculate Risk and Reward in Every Action*—turns that skill into day-to-day practice. Below is how I use it, where it fails, and how to keep it reliable in adversarial settings.

### The Decision Triad

- **Reward.** Start by stating exactly why the action exists. The *Reward* must be specific, measurable, and ranked. A product launch might target "first-page search ranking within 60 days"; a patrol might aim to "deny enemy use of Route Blue for 48 hours." Vague aims ("improve morale," "show strength") invite scope creep. My test is simple: What single statistic would convince a neutral observer we succeeded? Until that number is fixed, the rest is guesswork. This echoes Principle 2's demand to define victory precisely—just at a more nuanced level.

- **Cost.** Use a full ledger. Most teams count dollars; some count hours; few count the harder currencies—morale, political goodwill, options foreclosed. Those can dominate the result. A two-month sprint that burns out key staff can cut velocity for a year. I translate soft costs into the same unit the budget team respects, even if roughly (for example, one morale point ≈ two weeks of lost ramp-up). An imprecise estimate beats a blank cell.

- **Probability.** Treat it as a live variable. We tend to be optimistic when we think we control a domain and pessimistic when we do not. To steady the estimate, I use three habits: evidence ladders (weight strong, recent evidence above weak or stale evidence); range estimates (state a 90% interval, not a single point); and freshness clocks (timestamp each estimate and set a recalc trigger tied to the data's half-life—daily for cyber IOCs, monthly for consumer sentiment). If conditions change and we do not refresh *Probability*, the ledger is out of date. A rule I borrow from Douglas Hubbard helps: measurement is the art of reducing uncertainty, not declaring certainty. Ask, "What information would most shrink my uncertainty range?" Even a rough interval supports better choices than "no idea." Applied here, that mindset keeps *Probability* and *Cost* from being ignored just because they're hard to pin down.

## Guardrails That Keep the Math Honest

- **Iteration Cadence.** Calculations go stale. A sound day-one estimate can be wrong by day thirty after reinforcements arrive, supply chains slip, or competitors move. Schedule recalculation checkpoints in advance—daily in kinetic operations, weekly in product sprints, quarterly for infrastructure. Waiting until it "feels" stale is usually too late.

- **Heuristics.** Speed without panic comes from a short list of rehearsed rules. Under pressure, people pick the first acceptable option; preloaded, tested heuristics keep that choice sane. Pilots use "altitude, attitude, airspeed" to stabilize a failing aircraft in seconds. Traders use four-number position limits that force automatic scaling out of losses. Each team should write its own small set, test them in simulations, and debrief them after use. This is especially an important guardrail in conflict, where speed is a critical and ever-present factor.

- **Bias Countermeasures.** Fear overweights worst-case *Cost*; greed overweights best-case *Reward*; ego inflates *Probability*. I counter those with a rotating red-team role, premortems (assume failure and list causes), and kill-switch thresholds tied to loss or time that force an automatic pause and recalculation.

- **Risk Appetite and Portfolio.** Not every move should be low risk. If the *Reward* is existential—securing funding, stopping an invasion—I may accept a lower success probability, so long as total portfolio risk stays within tolerance. I write down the risk-appetite statement and review it on cadence to prevent the quiet pile-up of "reasonable" bets that add up to ruin. I think back to the Kodak case study from Principle 16. Kodak made

reasonable bets at every juncture and then lost everything
because they didn't take the larger risk.

- **Information Value.** Sometimes the best move is to pay to
learn. A reconnaissance flight or an A/B test that
materially shifts *Probability* can justify its own *Cost*. I price
learning deliberately, the way I'd price ammunition.
- **Exit Discipline.** Define exits before emotions take over. I
set both a stop-loss (cost threshold) and a take-profit
(reward threshold). Chasing a loser is obvious folly; so is
riding a winner past diminishing returns. The same ledger
that guards against rash entry also guards against
overextension.

Put simply: set *Reward*, count full *Cost*, estimate *Probability*—then
keep those three honest with cadence, heuristics, bias checks, portfolio bounds, paid learning, and precommitted exits. Keep that loop
intact and you earn Patton's "calculated." Let it slide and you drift
toward rash—and defeat.

### Positive Example — How *Moneyball* Repriced Baseball's Entire Marketplace

Beyond being a great example, this is one of my all-time favorite
movies. From the late 1990s into the early 2000s, MLB payroll gaps
ballooned toward three-to-one. The Yankees were spending in the
$110-plus-million range while my poor Tampa Bay Rays (then still the
Devil Rays) lived down near the $20 million basement. Oakland,
stuck in a decaying stadium without a regional TV empire, also sat
near the bottom in discretionary cash. Billy Beane and Paul
DePodesta faced an economics problem: they could not buy the same
inputs as the rich. They had to redefine inputs *and* redefine value.

For Oakland the *Reward* wasn't "look like a contender," it was
postseason access per dollar—wins, efficiently purchased. That
forced a simple causal chain: runs produce wins, and getting on base
produces a lot of runs. So they stopped paying for "five-tool optics"—

(*batting average, batting power, speed, defensive fielding*, and *arm strength*) which intuitively made sense but did not cleanly follow a causal chain to wins—and started paying for on-base events. The *Cost* side was a full ledger, not just salary. Roster spots, opportunity cost, and reputation risk all counted. That is why a converted catcher with nerve trouble (Scott Hatteberg) made sense: an inexpensive contract and a .370-ish OBP. It is why a submarine reliever (Chad Bradford) mattered: a ground-ball profile that smothered rallies late. It is why an aging slugger (David Justice) still had value: the bat wasn't what it used to be, but the ability to reach base *was*. None of that was pretty. All of it was cheap relative to the runs it bought.

The *Probability* work happened quietly. Projections were ranges that tightened with plate appearances; streaks were treated as noise until the data said otherwise. When the 2002 A's opened 20–26, talk radio lambasted the statistical approach to the game. The front office stayed confident in their probability math, let on-base rates normalize, watched run differential turn, and then rode the sustained surge: twenty straight wins from August 13 to September 4, 2002. They finished 103–59—the same win total as New York—at a fraction of the price per win. None of that required a checklist. It was the triad, kept honest by a few guardrails: a set cadence for recalculation, a handful of rehearsed heuristics (buy OBP, buy grounders late, don't pay for steals or RBIs without OBP), numbers that red-teamed scout lore without humiliating anyone, and clear exit discipline when a piece didn't fit. Oakland didn't outmuscle the market; it outcalculated it with refined risk-reward thinking.

### Negative Example — U.S. COVID-19 Response: Calculus Frozen by Politics

In early 2020 a novel coronavirus began to spread. On March 16, 2020, the White House launched "15 Days to Slow the Spread" of COVID-19 with the familiar "flatten the curve" chart: slow infections so hospitals don't overflow. As a first step, that aim was clear and time-boxed.

What needed to happen next was not more slogans but a logical and transparent way to calculate risk and reward in fighting the virus. Leaders had to show, in plain language, what we were trying to gain, what we were willing to pay, and how confident we were—then revisit those judgments on set dates. Instead, data was thin, agencies and states disagreed on which numbers mattered, and arguments hardened along political lines. The calculation never became common property, so the response never moved as one.

*Reward.* After the two-week surge plan, the *Reward* should have been restated for the next phase in a single sentence, and repeated everywhere: keep hospitals open and deaths down while keeping as much school and work going as possible. Each major rule—closures, reopenings, masks, travel—needed a short note explaining exactly how it served that goal. In practice, the target drifted. Some leaders shifted to daily case counts; others spoke as if the aim were "stop the virus." With no shared statement of benefit, agencies and governors pulled in different directions and the public could not tell what "winning" meant.

*Cost.* The *Cost* should have been shown on the same page as the aim—not just lost output, but learning loss, delayed medical care, mental-health strain, small-business closures, and the slow erosion of trust. Even rough ranges would have helped ("keeping schools remote this month likely means about this much remediation later"). Because those tradeoffs were rarely laid out together, many extensions looked reasonable on their own but added up to a larger, hidden bill. People felt the bill before they ever saw it.

*Probability.* Early *Probability* was wide; that's normal for a new disease. The job was to say what we believed now, what would change our mind, and when we would check again. Weak testing, shifting case definitions, and uneven reporting made estimates shaky; politics made it harder to admit uncertainty. Guidance flipped as new evidence came in, but often without a short "why this changed" and a promised review date. People were asked to change behavior without seeing how the odds had moved.

This is where process could have unified action. Imagine that every major decision came with a one-page public note: the *Reward* in one sentence; the key *Costs* we're accepting (health, school, economic, trust); today's *Probability* range and the data that would move it; plus the date we will revisit. Keep those notes in one place; update them on a regular schedule; keep the language plain. Even with noisy data and political heat, that method would have made disagreements honest and coordination easier. Instead, different actors chose different dials and defended them as identity, not analysis.

Intent wasn't the problem; the method was missing. Because the next-phase *Reward* wasn't stated and kept in view, the *Cost* wasn't priced in the open, and the *Probability* wasn't updated on a schedule, the calculation wandered—and the public felt the drift as whiplash. The bill was heavy: countless small businesses closed for good, workers fell out of the labor force, students lost months of learning, anxiety and depression spiked, routine care was delayed, and trust eroded just when consent mattered most. Had policymakers run a clear, shared *Reward–Cost–Probability* calculation and kept it current, actions would have been narrower and more coherent—schools reopened sooner with targeted mitigation, high-risk settings protected first, hospital capacity guarded without blanket measures, and rules retired once they no longer bought health. We wouldn't have made the virus easy, but we would have made the response purposeful—lower *Cost* for the same or better *Reward*, fewer reversals, and a public more willing to move together.

## Principles in Action — Reflection Questions

1. What single statistic will prove success, and is it visible team-wide? (*Reward*)
2. Which intangible costs (morale, optionality, regulatory friction) could outweigh line items, and how are we converting them into the same unit as cash? (*Cost*)
3. What is our current 90% *Probability* range, when was it last

refreshed, and what observation (test/recon) would shrink it fastest?

4. What is our Iteration Cadence, and who owns the recalculation?
5. What are our risk bounds and exits—the written Risk Appetite for this bet/portfolio, plus the precommitted stop-loss and take-profit?

## Sources

- The Atlantic. 2020. "How the Pandemic Defeated America." *The Atlantic,* April 2020.
- Ball, Deborah, and Daniel Levitt. 2023. "What Was the Cost of the Lockdowns?" *The Wall Street Journal,* March 10, 2023.
- Baseball Prospectus (Cot's Baseball Contracts). n.d. *Opening Day Payrolls, 2002.* Baseball Prospectus.
- Baseball-Reference. n.d. *Chad Bradford Statistics.* Sports Reference LLC.
- Baseball-Reference. n.d. *David Justice Statistics.* Sports Reference LLC.
- Baseball-Reference. n.d. *Oakland Athletics 2002 Game Log.* Sports Reference LLC.
- Baseball-Reference. n.d. *Scott Hatteberg Statistics.* Sports Reference LLC.
- Baseball-Reference. n.d. *2002 Oakland Athletics Team Page/Statistics.* Sports Reference LLC.
- Bendavid, Eran, Christopher Oh, Jay Bhattacharya, and John P. A. Ioannidis. 2021. "Assessing Mandatory Stay-at-Home and Business Closure Effects on the Spread of COVID-19." *European Journal of Clinical Investigation* 51 (4): e13484. https://doi.org/10.1111/eci.13484
- BizTech Magazine. 2017. *How Data Analytics Is Changing Baseball.* Arlington, VA: BizTech Magazine.

- Brauner, Jan M., et al. 2021. "Inferring the Effectiveness of Government Interventions against COVID-19." *Science* 371 (6531): eabd9338. https://doi.org/10.1126/science.abd9338
- Centers for Disease Control and Prevention. 2020. "Recommendation Regarding the Use of Cloth Face Coverings to Help Slow the Spread of COVID-19." Atlanta, GA: CDC, April 3, 2020.
- Centers for Disease Control and Prevention. 2021. "Mental Health Surveillance—United States, 2020–2021." *MMWR Supplement.* Atlanta, GA: CDC.
- Congressional Budget Office. 2022. *Budgetary Effects of the 2020 CARES Act.* Washington, DC: CBO.
- ESPN. 2015. *The Great Analytics Rankings.* Bristol, CT: ESPN.
- Gardner, Howard. 1983. *Frames of Mind: The Theory of Multiple Intelligences.* New York: Basic Books.
- Gottfredson, Linda S. 1997. "Why g Matters: The Complexity of Everyday Life." *Intelligence* 24 (1): 79–132. https://doi.org/10.1016/S0160-2896(97)90014-3
- Haug, Nils, et al. 2020. "Ranking the Effectiveness of Worldwide COVID-19 Government Interventions." *Nature Human Behaviour* 4: 1303–1312. https://doi.org/10.1038/s41562-020-01009-0
- Howard, Jeremy, Austin Huang, Zeynep Tufekci, and Trisha Greenhalgh. 2021. "An Evidence Review of Face Masks Against COVID-19." *Proceedings of the National Academy of Sciences* 118 (4): e2014564118. https://doi.org/10.1073/pnas.2014564118
- Hubbard, Douglas W. 2014. *How to Measure Anything: Finding the Value of Intangibles in Business.* 3rd ed. Hoboken, NJ: Wiley.
- Lewis, Michael. 2003. *Moneyball: The Art of Winning an Unfair Game.* New York: W. W. Norton.
- McKinsey & Company. 2021. *COVID-19 and Student Learning in the United States: The Hurt Could Last a Lifetime.*

- Major League Baseball. n.d. *Qualifying Offer.* New York: MLB Advanced Media.
- Patton, George S. 1947. *War as I Knew It.* Boston: Houghton Mifflin.
- Pew Research Center. 2022. *Americans' Trust in Scientists and Other Groups Declines.* Washington, DC: Pew Research Center.
- Sternberg, Robert J. 1988. *The Triarchic Mind: A New Theory of Human Intelligence.* New York: Viking.
- U.S. Bureau of Economic Analysis. 2020. *Gross Domestic Product, 2nd Quarter 2020 (Second Estimate) and Corporate Profits, 2nd Quarter 2020 (Preliminary Estimate).* Washington, DC: BEA.
- The White House. 2020. *The President's Coronavirus Guidelines for America: 15 Days to Slow the Spread.* Washington, DC: The White House, March 16, 2020.

# PRINCIPLE 25
## MAINTAIN COMPOSURE UNDER PRESSURE

"We must never lack calmness and firmness, which are so hard to preserve in time of war. Without them the most brilliant qualities of mind are wasted." — Carl von Clausewitz

P ressure arrives before any adversary does. A siren, an unexpected headline, a blinking cockpit panel—each can dump adrenaline into the bloodstream. The surge is useful for sprinting or lifting wreckage, but it sabotages judgment. As arousal rises, the prefrontal cortex can cede executive control to limbic circuits; as heart rate spikes, peripheral vision narrows and hearing tightens; memory for multi-step procedures fractures into fragments. The brain trades foresight for reflex. Clausewitz called the gulf between a tidy plan and physiological reality *friction*. Leaders who prevail under friction do not possess superhuman calm; they cultivate it the way one builds endurance—through repeated stress, structured reflection, and a clear destination that orients thought when chaos scrambles attention.

Mission clarity is the first inoculant. In every crisis room I've sat in, someone eventually asked, "What is the mission right now?" When we could answer in a sentence—rescue the hostage alive, safe-

guard the election database, evacuate in under twenty-four hours—
voices steadied and tasks aligned. When the mission was fuzzy,
cortisol rose and side debates bloomed. This is yet another nod to
*Principle 2: Define Victory with Precision.* Johnson & Johnson's Tylenol
response is textbook: the company credo—placing patients first—
acted as a North Star that kept lawyers, engineers, and marketers
from spinning into self-preserving factions when disaster struck. As
we will see in the positive example below, the calm tone of James
Burke's briefings drew power from that fixed point; he could speak
plainly because he already knew whose welfare came first.

*Dune* captures the intimate version of this discipline with the
Bene Gesserit *Litany Against Fear*, which involves talking oneself
down from the illogical mindset that accompanies acute stress. It isn't
neuroscience, but the litany mirrors modern stress work: name the
fear, breathe through its onset, visualize its exit. Those steps down-
regulate the amygdala and return control to deliberate systems. I've
leaned on that idea more times than I care to admit: fear is the mind-
killer not because danger isn't real, but because unmanaged arousal
hijacks the very circuitry we need to think.

Composure is therefore both biochemical and narrative. We train
the biochemical side with stress-inoculation drills: SWAT units
clearing rooms with flash-bangs detonating inches away; software
engineers paged at 3:02 a.m. to diagnose deliberately broken
microservices; basketball players who simulate end-game fatigue by
sprinting before free-throw practice. Each controlled exposure lays
myelin on the pathways that must work when oxygen debt or sensory
overload arrives for real. This is why Brazilian Jiu-Jitsu (BJJ) classes
devote most sessions to rolling—unscripted grappling at near-full
effort. Students tap out thousands of times; each small failure inocu-
lates against panic later.

The narrative side relies on precommitted values and a shared
mental model because stress clusters the mind around whatever
story is most available. If the story is coherent—"Patients first," "Land
the aircraft intact," "Keep the hostage breathing"—decisions flow
toward the basin those words define. If the story is vague or conflict-

ing, the basin splits and latency stretches. Greg Norman's 1996 Masters collapse began the moment his inner monologue shifted from process—breathe, visualize, execute—to outcome. "Don't blow this lead!" Anxiety tightened muscles and shortened his routine, and fine-motor error followed.

Leaders amplify whichever story they embody. Research on the human mirror system shows that people quickly echo others' states; calm or agitation spreads fast through a room. On a submarine bridge, the captain's voice sets a metronome for a dozen operators; at a crisis podium, a CEO's cadence sets investor breathing almost in step. That contagion makes deliberate self-regulation part of the job. A mentor once tapped his watch before a high-stakes briefing I had to give and said, "Three-second breaths, or they'll breathe twice as fast listening to you." We practiced until it was automatic. The room only saw the calm.

Training for pressure isn't about flawless execution; it's about composure plus adaptive action. Checklists free working memory but must be short enough to survive tachycardia. People won't remember long, detailed processes in a moment of intense panic. It's therefore important that after-action reviews (AARs) convert fresh, adrenaline-tagged memories into protocol before decay dulls the lesson.

Composure is also a type of resource that must be replenished. Organizations that skip decompression efforts pay later. Chronic stress effects often surface weeks after a crisis, hollowing teams just as normal operations resume. Militaries counter this with mandated reset cycles; companies too often congratulate themselves on surviving an end-of-year sales push while burnout germinates. A true composure culture covers the whole arc: spike, sustain, pivot, decompress, reflect.

The through-line is simple: you do not rise to the occasion; you fall to the level of your training and your narrative clarity. A mantra, a credo, a one-sentence mission—these are not décor. They script the mind's next line when stress would otherwise improvise disaster. With that frame, the Tylenol recall and the Masters collapse read as opposite ends of a composure spectrum: one anchored by a credo that absorbed shock, the other untethered by a whisper of doubt.

## Positive Example — Johnson & Johnson's Tylenol Crisis Response

On the morning of September 29, 1982, 12-year-old Mary Kellerman of Elk Grove Village, Illinois, swallowed an Extra-Strength Tylenol and died within hours. By nightfall, six more Chicago-area residents lay dead; all had taken cyanide-laced capsules from tampered Tylenol bottles. At the time, Tylenol controlled roughly a third of the U.S. over-the-counter analgesic market and contributed roughly one-fifth of Johnson & Johnson's profit. Reporters and FDA agents camped outside headquarters in New Brunswick, New Jersey, poised to write obituaries for the brand and, possibly, the entire corporation.

James Burke, J&J's chairman, convened his crisis team the next morning. The credo hanging in every hallway—"The needs of the patient come first"—shifted from mere wall art to meaningful metric. Within 48 hours the company halted all Tylenol production and advertising, undertaking a first-of-its-kind nationwide recall for a major OTC brand. Trucks fanned out across the country, retrieving 31 million bottles—costing an estimated $100 million.

Press briefings adopted unusual transparency. Spokesperson Larry Foster avoided legal jargon, explaining capsule lot numbers, cyanide toxicity, and ongoing cooperation with the FDA in language a high-school student could parse. Questions about lawsuits landed on camera, and Foster answered without defensiveness: "We will do whatever it takes to protect consumers." The plain speech steadied public nerves; panic buying subsided within a weekend.

Behind the microphones, engineers raced to prototype triple-tamper seals: glued flaps, foil inner seals, and plastic shrink bands. One executive later likened the crash retooling to an 'Apollo 13 in reverse'—a save-the-company sprint. Factories retooled quickly. By November 11, 1982, Tylenol reentered shelves in triple-sealed packaging, supported by national ads and $2.50 coupons that explained the new safety measures. Market share rebounded from about 8 percent immediately after the poisonings to roughly 24 percent by early 1983. Within 18 months Tylenol reclaimed its precrisis dominance.

Composure animated every layer of the response. Burke's calm press appearances signaled confidence upstream to suppliers and downstream to consumers. Executives resisted legal counsel urging tepid action—fear of liability yields defensive silence; composure under mission clarity yields decisive transparency. Employees, watching leadership accept financial pain without blame games, doubled overnight shifts to retrofit lines. The Tylenol case still endures because it transformed catastrophe into trust, showing how practiced calm paired with swift, values-aligned action can turn public terror into loyalty stronger than before.

### Negative Example — Greg Norman's Masters Collapse

Greg Norman arrived at Augusta National on Sunday, April 14, 1996, holding a six-stroke lead after 54 holes—a Masters record. He had already spent 331 weeks ranked world number one, earned the nickname "Great White Shark," and looked poised to slip into his first green jacket. Gallery whispers suggested only weather could stop him, but pressure proved the fatal opponent.

The opening tee shot sailed, but Norman left his approach on the first hole short, chunked a chip, and settled for bogey. On the par-3 sixth, his 6-iron ballooned right into a bunker; another bogey. Meanwhile, playing partner Nick Faldo birdied two holes. Lead down to three.

Physiology edged in. Norman's tempo quickened as heart rate spiked. Sports psychologists later noted his preshot routine shrank by about three seconds—small on paper, huge in a game built on rhythm, and very telling. At Amen Corner's 155-yard 12th, Norman aimed safely center but pushed the swing, the ball dunking into Rae's Creek. Faldo birdied. The leaderboard flipped for the first time all week.

Walking to the 13th tee, Norman told caddie Tony Navarro, "I just don't have it." Self-talk turned traitorous; cortisol narrowed vision. A pulled drive and water lay-up led to another bogey. On 16, his 7-iron came up 10 feet short and spun back into the pond. Double bogey. By the 18th green, the Shark needed an eagle just to tie; he carded a final-round 78, six strokes over par, losing by five.

Analysts pointed to mechanical flaws—hips clearing too early—but biomechanics were symptoms. The root was composure fracture. Under Sunday pressure, Norman's feedback loop spiraled: each miss fed doubt; doubt shortened routine; shortened routine invited another miss. Faldo kept a neutral expression, breathing through pursed lips, expanding his own decision bandwidth while Norman's collapsed. Where Johnson & Johnson expanded trust by remaining calm and decisive under existential threat, Norman's unspooling showed how quickly mastery erodes when composure falters.

## Principles in Action — Reflection Questions

1. When high stakes hit, do we default to trained breathing, checklists, and value anchors—or to improvisation?
2. How do leaders model "control-room voice" so the team's cortisol follows calm rather than panic?
3. Have our drills pushed heart rates and decision-latency metrics to crisis levels, or just rehearsed theory?
4. Do we decompress teams postcrisis to prevent delayed burnout once adrenaline fades?
5. What single value statement guides split-second choices, letting calm flow from principle instead of propaganda?

## Sources

- Beilock, Sian. 2010. *Choke: What the Secrets of the Brain Reveal About Getting It Right When You Have To*. New York: Free Press.

- Bloomberg Businessweek. 2012. "Tylenol Survivor: How J&J Triumphed After Tragedy." *Bloomberg Businessweek,* September 2012.
- Burson-Marsteller. 1984. *Tylenol Crisis Case Study.* Public Relations Society of America.
- Clausewitz, Carl von. 1984. *On War.* Translated by Michael Howard and Peter Paret. Princeton, NJ: Princeton University Press.
- Food and Drug Administration. 1989. "Tamper-Resistant Packaging Regulations for OTC Human Drug Products; Final Rule." February 2, 1989.
- Futterman, Matthew. 2011. "Masters Meltdown Still Haunts Norman." *The Wall Street Journal,* April 8, 2011.
- Gawande, Atul. 2010. *The Checklist Manifesto: How to Get Things Right.* New York: Metropolitan Books.
- Grossman, Dave, and Loren W. Christensen. 2004. *On Combat: The Psychology and Physiology of Deadly Conflict in War and in Peace.* Belleville, IL: PPCT Research Publications.
- Herbert, Frank. 1965. *Dune.* Philadelphia: Chilton Books.
- Iacoboni, Marco, et al. 2001. "Neural Mechanisms of Human Imitation and Interpersonal Interaction." *Journal of Neuroscience* 21 (19): 7767–73. https://doi.org/10.1523/JNEUROSCI.21-19-07767.2001
- McGonigal, Kelly, et al. 2016. "Breath-Paced Practices Reduce Heart-Rate Reactivity Under Acute Stress." Stanford Center for Compassion and Altruism Research and Education.
- Sports Illustrated. 1996. "Faldo Calmly Cruises as Norman Crumbles." *Sports Illustrated,* April 22, 1996.

# PRINCIPLE 26

## NURTURE YOUR CRITICAL RESOURCES

*"Take care of your ship, and she'll take care of you."* — Traditional Naval Maxim

This timeless naval wisdom reveals a deep truth about any adversarial contest. A ship is more than a vessel; it is a living system of people and technology. Victory in a storm belongs not just to the ship with the biggest guns, but to the one whose crew is best cared for and whose hull is sound. To nurture your critical resources is to practice this logic of pragmatic steward-ship. It asserts that decisive advantage flows from the unglamorous, daily routines that keep people empowered, equipment functional, and morale intact. If *Principle 3* is about *building* the team and tooling, then *Principle 26* is about *sustaining* it.

This is the work of the gardener, not the gambler. Where other principles focus on the decisive blow, this one focuses on the health of the arm that swings the sword. The modern temptation is to treat this work as a backstage concern, separate from "real" action. That divide is fatal. An adversary who can predict when your organization will be depleted does not need superior firepower; they need only patience.

## Three Dimensions of Nurturing

1. **Nurturing People.** Human beings are not machine parts; they are organic systems that require more than just fuel. They burn calories, cortisol, and conviction. Nurturing people means designing cycles of exertion and recovery. But more deeply, it means fostering the psychological soil in which they thrive: trust, empowerment, psychological safety, and a clear sense of purpose. Morale is a renewable resource, but only if the conditions for its renewal are cultivated.

2. **Nurturing Equipment and Infrastructure.** From rifles to routers, every tool decays. Nurturing equipment is about more than just maintenance; it's about a culture of stewardship. It requires preventative upkeep that is honored even when emergencies tempt postponement. It demands anticipatory investment—ordering spare parts before the shortage. And crucially, it requires *right-fitting the tool to the task*, ensuring that the equipment is not just functional, but appropriate for the environment in which it must perform.

3. **Nurturing Intangible Capital.** Data integrity, brand reputation, and political goodwill are operating fuel. A police department that burns community trust loses informants; a tech company that ships buggy code spends down customer patience. Treat these reservoirs as stocks: measure them, protect them, and deliberately replenish them.

## Nurturing as an Offensive Weapon

Counterintuitively, this discipline is not meek conservatism; it is a vital component of an aggressive posture. A well-nurtured force can seize opportunities that would break a brittle one. By running round-the-clock, rotating shifts that kept cryptanalysts rested and decision-ready, Bletchley Park cracked Enigma at a pace few anticipated. Each example treats endurance not as insurance, but as a means to seize and hold the initiative—because the side that is healthiest when opportunity appears often dictates the next move. Endurance wasn't only man-hours and spare parts; it was *intangible capital*—the credibility that won cooperation, the data discipline that kept signals clean, and the trust that let leaders move fast without relitigating every call.

### Positive Example — The U.S. Navy's Nuclear Power Program

Perhaps no organization in history better embodies the principle of nurturing critical resources than the U.S. Navy's Nuclear Power Program. Forged by the famously demanding Admiral Hyman G. Rickover, the program faced an unforgiving adversary: the physics of nuclear fission. In the high-stakes environment of the Cold War, a single failure of people or equipment could lead to a catastrophic reactor accident and a national disaster. The program's survival and dominance were built on the ultimate expression of the naval maxim: *they took care of their ship.*

The selection and training for the naval nuclear program became legendary. Only top academic performers were considered, and they were subjected to a grueling training pipeline designed not just to teach them engineering, but to instill a culture of absolute intellectual honesty and procedural compliance. Rickover personally interviewed virtually every officer candidate for the nuclear program for decades. This intense process created an elite cadre of operators who were deeply empowered. Any watch officer was expected—and required—to shut down the reactor if something looked wrong, without fear of reprisal. This is the ultimate form of empowerment. The program was adept at *Nurturing People.*

The program's approach to technology was equally rigorous. Every component was designed and built to the highest possible standard of safety and reliability. Maintenance was not a secondary task; it was a core operational ritual. Procedures were followed to the letter, and any deviation was intensely scrutinized. The program's practice of *Nurturing Equipment and Infrastructure* was cemented by its refusal to cut corners, ever.

The result of this deep, cultural commitment to nurturing resources is one of the most remarkable safety and performance records in history. In more than 70 years of operation, U.S. Naval Reactors reports no reactor accidents—no events causing harm to the public or the environment. This perfect record is not luck. It is the direct outcome of a system designed to nurture its people and equipment with relentless, daily discipline. This unwavering reliability, in turn, became a massive strategic advantage, allowing the U.S. submarine force to operate with a level of confidence and dominance its adversaries could never match. This reliability was also a way of *Nurturing Intangible Capital.* It led to Congressional and public trust in the program, which in turn protected budgets and freedom to operate.

### Negative Example — The Space Shuttle *Challenger* Disaster

On the freezing morning of January 28, 1986, the Space Shuttle *Challenger* disintegrated 73 seconds after liftoff, killing all seven crew members. The technical cause was a joint failure in a solid rocket booster. But the true, strategic cause was a systemic *failure to nurture* the critical resources of people and equipment—a failure to "take care of the ship."

The night before the launch, engineers from the contractor Morton-Thiokol pleaded with NASA managers for a delay. They presented data and recommended not launching below 53°F, the coldest prior flight, because the booster O-rings stiffen in cold and could fail. After management pressure during the teleconference, the contractor reversed its no-launch recommendation. The decision path elevated schedule over engineering judgment. Their expertise was the most critical human resource available at that moment. But NASA's culture, driven by intense schedule pressure, failed to nurture or empower them. The engineers' warnings were overruled by managers who were not technical experts. Instead of being valued, their dissent was treated as an obstacle. This was the opposite of the Rickover model; it was a culture that actively suppressed its most vital human feedback in a blind drive for mission completion. Compounding this, NASA's independent safety function had little standing in launch decisions—what the Rogers Commission later described as a "silent" safety program. This was the opposite of *Nurturing People.*

The O-rings themselves were a resource that had been dangerously neglected. For years, engineers had documented evidence of heat damage and erosion on the O-rings on previous flights. Instead of treating these signals as a dire warning and nurturing the hardware—by redesigning the joint or refusing to fly until it was fixed—NASA's management began to accept the flaws as a "normal" and acceptable risk. This "normalization of deviance," as sociologist Diane Vaughan termed it, is the antithesis of nurturing a critical component. They were allowing a vital piece of their machine to degrade in plain sight. This violated what should have been a core element of the program: *Nurturing Equipment and Infrastructure.*

The *Challenger* tragedy is a powerful negative lesson. NASA had brilliant people and extraordinary technology. But its culture failed to nurture them with empowerment. It ignored the warnings of its experts and tolerated the degradation of its equipment. This created a strategic vulnerability that the unforgiving environment of physics exploited with catastrophic results. The loss also had the opposite effect of *Nurturing Intangible Capital*—internal trust, public confidence, and political support shattered—making future safety reforms more expensive and slower to win.

## Principles in Action — Reflection Questions

1. Is our 'ship' sound? Do we regularly inspect our crew's

morale and training, our equipment's integrity, and our intangible capital (trust, data quality, reputation)?

2. Are our people—especially our technical experts—truly empowered to sound the alarm when they see a flaw in the hull, without fear of reprisal?

3. Does our culture reward the unglamorous, preventative work of maintenance and training, or do we only celebrate last-minute heroics?

4. Are we right-fitting our tools for the job, or are we asking our "ship" and crew to perform in conditions they were not designed for?

5. Does our budget reflect a commitment to nurturing our assets through continuous upkeep, or do we only fund repairs after a storm has already hit?

**Sources**

- Edmondson, Amy C. 2018. *The Fearless Organization: Creating Psychological Safety in the Workplace for Learning, Innovation, and Growth.* Hoboken, NJ: Wiley.
- NASA. 1986. *Report of the Presidential Commission on the Space Shuttle Challenger Accident.* Washington, DC: U.S. Government Printing Office.
- Rockwell, Theodore. 1992. *The Rickover Effect: How One Man Made a Difference.* Annapolis, MD: Naval Institute Press.
- U.S. Department of Energy and Department of the Navy. 2019. *Occupational Radiation Exposure from U.S. Naval Nuclear Propulsion Plants and Their Support Facilities.* Washington, DC: Naval Reactors/DOE.
- Vaughan, Diane. 1996. *The Challenger Launch Decision: Risky Technology, Culture, and Deviance at NASA.* Chicago: University of Chicago Press.

# PRINCIPLE 27

## PACE YOUR ACTIONS FOR SUSTAINABILITY

*"Slow and steady wins the race."* — Aesop

In adversarial dynamics, speed is important—and seductive—but it is not sovereign. A burst of intensity may win a moment. Only endurance wins the struggle. To *Pace Your Actions for Sustainability* is to understand that every system—human, organizational, biological—operates on finite reserves. It is a discipline of calibration: knowing not just how much force to apply, but *when* and *for how long*. Misjudging the duration of a conflict leads to collapse, not from enemy strikes, but from internal depletion.

Sustainability is not slowness. It is *intentionality*. It means aligning effort with purpose, rhythm with terrain, and intensity with anticipated length. A well-timed sprint can change the shape of an engagement, but a life lived in sprint mode ends in burnout. This principle demands that we resist the instinct to escalate reflexively and instead cultivate the foresight to conserve energy for the moment that matters most.

This principle is the direct complement to Principle 26, *Nurture Your Critical Resources*. Principle 26 is about the fundamental health of the engine—the daily maintenance, the quality of the parts, and the skill of the mechanic. It represents doing the oil changes and tune-ups. This principle, Principle 27, is about how you *drive* that engine during the race: knowing when to redline on the straightaways and when to conserve fuel in the corners. One is about the *readiness* of the machine; the other is about the *rhythm* of its performance.

## The Wisdom of the Tortoise

Few stories capture this principle more enduringly than Aesop's fable of *The Tortoise and the Hare*. Though deceptively simple, the tale encodes a deep truth about adversarial pacing. The hare begins with explosive advantage—speed that far outstrips his opponent. But his burst is mismatched to the length of the race. He coasts, pauses, and ultimately mismanages his effort. The tortoise, slower by every metric, wins through constancy. He applies the precise amount of energy needed to maintain motion from start to finish. His pace is sustainable, and in the end, decisive.

That victory is not accidental. It is structural. The tortoise recognizes that the engagement will be long. He does not need to match his adversary's speed—only to move at a rate that can endure. In doing so, he turns his limitation into a strength.

The same pattern emerges in conflict after conflict. Whether in military campaigns, market competitions, sports tournaments, or psychological battles, those who calibrate their output to the full arc of the engagement consistently outperform those who burn through their strength in its early phases.

## The Four Currencies of Energy

To pace effectively, one must first understand what is being spent. "Energy" is not a metaphor. It is a measurable force drawn from multiple domains, and it decays with misuse. In UTAD, we frame this through four interdependent currencies:

- **Physical Energy:** The stamina of people, equipment, and physical systems. Depletion here results in fatigue, breakdown, or collapse.
- **Cognitive Energy:** The capacity for clear thought, pattern recognition, and disciplined decision-making. Under strain, this currency diminishes into tunnel vision or error-prone reactivity.
- **Emotional Energy:** Morale, cohesion, conviction, and resilience. It fuels perseverance in adversity, but once depleted, its absence destabilizes intent.
- **Systemic Energy:** The operational efficiency of an organization. As internal friction rises—through misalignment, overload, or bureaucracy—energy is lost to heat, not motion.

Strategic pacing means managing all four simultaneously. When any one collapses, the rest soon follow.

## The Trap of the Sprint

The most common failure of pacing is not under-exertion—it is over-exertion at the wrong time. In fast-moving or competitive environments, the temptation to "go all in" early is powerful. We equate action with progress, urgency with success. But sprinting through a marathon is a strategy of self-destruction.

This trap is especially common in systems driven by ambition or fear. Consider the startup that mistakes continuous overwork for momentum. Short-term gains mask long-term corrosion. Founders pride themselves on their "grind," but beneath the surface, their people erode—burning through cognitive and emotional reserves that cannot be replaced with capital alone. What looks like initial victory often ends in collapse.

Adversaries who operate in this mode become easy to outlast. Their exhaustion becomes your terrain.

*An important consideration:* Simon Sinek describes one kind of long-duration engagement as an "infinite game"—one in which the goal is not to win, but to stay in the game. In an infinite game, there aren't winners and losers in the usual sense—only those who are ahead and those who are behind. In these environments—corporate ecosystems, geopolitical contests, activist movements, even reputation battles—the adversarial dynamic is not defined by finality. There is no finish line. The primary strategic risk is not losing to an opponent, but exhausting oneself into irrelevance. In such games, pacing is not just a smart tactic—it is the core of survivability. Those who treat the infinite as finite, who sprint as if there's a decisive end in sight, burn themselves out before the real conflict even matures.

## The Strategic Pause

Central to sustainable pacing is the capacity to pause—deliberately, not passively. Strategic pauses are intervals designed for recovery, reorientation, and recalibration. In cultures that idolize constant motion, choosing to pause can appear weak or wasteful. In reality, it is an advanced form of control.

To pause well requires discipline. It is the moment to observe what has changed (Principle 17), to orient toward new conditions, and to update the course of action. It is not an absence of will—it is a reinforcement of it. The most sophisticated systems do not sprint until collapse; they cycle between focused action and structured recovery, treating each as equally strategic.

## The Leader as Pacemaker

Every team inherits its tempo from the top. Leaders set not only goals but pace. An erratic leader creates erratic systems. One who responds to every stimulus with immediate escalation builds teams that do the same. The result is internal exhaustion long before external threats are neutralized.

Strategic leadership requires resisting this impulse. It involves saying no to the unnecessary, enforcing rest even under pressure, and modeling composure. Leaders must monitor the burn rate across all four energy domains and protect against systemic overload. Sometimes this means pulling back when the culture demands push. Sometimes it means refusing to answer chaos with chaos.

To pace an organization is to ensure it remains whole long enough to finish what it started.

## Positive Example — The San Antonio Spurs and the Art of Load Management

When Coach Gregg Popovich sat Tim Duncan, Tony Parker, Manu Ginóbili, and Danny Green for a nationally televised game on November 29, 2012, commentators called it cheating the fans. The NBA fined the Spurs $250,000 for sitting healthy stars. Popovich paid the fine with a shrug because he was playing a longer game. By the early 2010s, San Antonio became an early adopter of team-wide workload management—scheduled rest on back-to-backs, lighter practice loads for veterans, and deliberate minutes caps—prioritizing playoff readiness over regular-season optics. The data showed that even superstar bodies lose a step by March if December workloads are not throttled.

Popovich institutionalized this type of load management. Veterans skipped back-to-back road games. Practices and minutes were calibrated to manage cumulative fatigue. Younger players logged surprise 30-minute nights so the bench gained playoff seasoning. Critics argued the strategy would sacrifice seeding, but the Spurs reached the 2013 NBA Finals anyway and returned in 2014 to dismantle the Miami Heat in five games, winning by an average margin of 14 points—the largest in NBA Finals history.

Behind the box scores lay ruthless pacing math. By limiting Ginóbili to 22.8 minutes per game, Popovich preserved his downhill burst for crunch time. Duncan's defensive rating stayed elite because his knees were not ground to dust in January. The Spurs converted rest into efficiency, and efficiency into rings. Popovich's methods soon spread league-wide, proving that judicious energy expenditure today multiplies combat power tomorrow.

## Negative Example — Theranos and the Illusion of Sprint-Mode Progress

Theranos, once a Silicon Valley darling, is now a case study in how the illusion of early velocity can disguise fatal structural weakness. Founded in 2003 by Elizabeth Holmes, the company promised to revolutionize blood testing with a device that could run hundreds of diagnostics on a single drop of blood. Investors and media celebrated its speed: rapid valuations, dazzling board appointments, and partnerships with major pharmacies followed in short order.

But beneath the surface, Theranos was not pacing itself for sustainability—it was sprinting through credibility on fumes. Its core technology did not work, and rather than slow down to refine it, Holmes doubled down on rapid expansion. Labs were rushed into deployment before validation. Employees were pushed to meet deadlines that physics and biology would not accommodate. Internal dissent was silenced. Engineers and scientists burned out. What looked like relentless progress was, in reality, reckless depletion of the system's core energies—physical, cognitive, emotional, and systemic.

The damage was not confined to Theranos alone. Patients received inaccurate test results. Regulators were misled. Investor confidence in health tech plummeted. And when the company collapsed in 2018—after exposés by whistleblowers and investigative reporting—it left behind not only legal consequences for its founder, but a cautionary tale for the entire innovation ecosystem.

Theranos did not lose to a better-funded rival. It lost to its own refusal to pace. Its sprint toward market dominance ignored the basic requirements of sustainability: proof, process, and pause. It is a modern parable of what happens when leaders mistake velocity for viability.

## Principles in Action — Reflection Questions

1. Are we matching our operational tempo to the true expected length of this conflict—or burning out too soon?
2. What signals tell us when to push and when to pause—and are we respecting them?
3. Which energy domains in our system (physical, cognitive, emotional, systemic) are most depleted—and what's our plan to restore them?
4. Are we building structured recovery into our rhythm, or reacting continuously without rest or recalibration?
5. As leaders, are we modeling calm and sustainable pacing —or driving the system into exhaustion by example?

## Sources

- Aesop. n.d. "The Tortoise and the Hare." In *Aesop's Fables.*
- Basketball-Reference.com. n.d. "Manu Ginóbili." Sports Reference LLC. http://www.basketball-reference.com/players/g/ginobma01.html
- Carreyrou, John. 2015. "Hot Startup Theranos Has Struggled with Its Blood-Test Technology." *The Wall Street Journal,* October 16, 2015. http://www.wsj.com/articles/theranos-has-struggled-with-blood-tests-1444881901
- Centers for Medicare & Medicaid Services (CMS). 2016. "Notice of Imposition of Sanctions (Theranos, Inc., Newark, CA)." July 7, 2016. http://www.wsj.com/public/resources/documents/r_Theranos_Inc_CMS_07-07-2016_Letter.pdf
- Department of Justice (DOJ). 2022. "Elizabeth Holmes Sentenced to More Than Eleven Years for Defrauding Investors." Press release, U.S. Attorney's Office, Northern District of California, November 18, 2022. http://www.

justice.gov/usao-ndca/pr/elizabeth-holmes-sentenced-more-11-years-defrauding-theranos-investors-hundreds

- ESPN Stats & Information. 2014. "Spurs Historically Dominant in 2014 Finals." *ESPN,* June 16, 2014. http://www.espn.com/blog/statsinfo/post/_/id/91567/spurs-historically-dominant-in-2014-finals
- Goldsberry, Kirk. 2019. *Sprawlball: A Visual Tour of the New Era of the NBA.* Boston: Houghton Mifflin Harcourt.
- National Basketball Association. 2012. "NBA Fines San Antonio Spurs $250,000 for Resting Players." *NBA Communications,* November 30, 2012. http://pr.nba.com/spurs-fined-250000-nba/
- Sinek, Simon. 2019. *The Infinite Game.* New York: Portfolio/Penguin.
- U.S. Securities and Exchange Commission (SEC). 2018. "SEC Charges Theranos, CEO Elizabeth Holmes, and Former President Ramesh 'Sunny' Balwani with Massive Fraud." Press Release 2018-41, March 14, 2018. http://www.sec.gov/newsroom/press-releases/2018-41

# PRINCIPLE 28
## UPHOLD YOUR INTEGRITY

*"For what will it profit a man if he gains the whole world, and loses his own soul?"* — Mark 8:36

Among the principles in this framework, few are more difficult to practice—or more important to preserve—than this one. *Uphold Your Integrity* means maintaining your own internal principles under pressure, refusing to sacrifice your identity, values, or moral compass even in the most hostile or consequential engagements. While most principles focus on how to win, this one addresses a different axis of survival: how to remain intact as yourself, even in the face of intense adversarial strain. As Nietzsche warned, "He who fights with monsters should take care that he himself does not become a monster."

Some battles are fought for territory, resources, votes, or market share. But the deeper contest runs through your own decision-making core. The risk is not only that the adversary might defeat you but that, in the process of resisting, you might become someone unrecognizable. This principle is about seeing that danger, naming it, and holding your ground against it.

Integrity is not a cosmetic layer. It is the structural architecture of strategic identity. Abandon it, and you may win the engagement—but lose the thing that made your victory meaningful in the first place.

## Integrity as a Strategic Asset

Integrity is often described in moral or personal terms—virtue, honor, righteousness. Those categories matter. But in adversarial dynamics, integrity is not just a virtue. It is also a *strategic asset*.

An entity known for internal consistency, ethical steadiness, and value-aligned behavior develops trust-based durability in the field. It becomes easier to coordinate with, easier to predict, easier to follow. Integrity reduces internal friction and speeds up decision-making, because people believe what is said, and trust that future actions will reflect present commitments. Subordinates act with confidence. Partners stay loyal. Even adversaries are more cautious.

This kind of moral consistency has gravitational pull. Allies rally more readily. Bystanders offer support. A clearly principled actor is harder to isolate, harder to mischaracterize, and harder to divide.

The benefits are structural:

- **Cohesion** (Principle 5) is strengthened.
- **Stakeholder alignment** (Principle 10) is stabilized.
- **Alliance credibility** (Principle 6) is enhanced.
- **Long-game adaptability** (Principle 17) is reinforced.

In some domains—politics, business, counterinsurgency, or diplomacy—where reputation is a kind of operating capital, integrity isn't just a moral choice. It's strategic positioning.

## Moral Collapse as Strategic Collapse

We often think of collapse in adversarial terms as external: overextension, resource depletion, exhaustion. But collapse can also happen internally, through a slow erosion of ethical boundaries until the system loses coherence from within.

This process is rarely sudden. It begins with exceptions. A small cut corner, justified by urgency. A truth withheld "just this once." A partner overlooked in pursuit of expediency. No single act breaks the system. But each one shifts the center of gravity away from the values that held the enterprise together. Over time, the mission becomes hollow. The signals get crossed. The identity fragments.

That's why this principle functions as a constraint on all others. You can master deception, pressure, momentum, and even psychological warfare, but without integrity as the boundary, you risk undermining your own long-term viability.

The most devastating failure is not being beaten. It is beating yourself.

## Defining Your Non-Negotiables

Integrity in conflict is not an abstract feeling. It's a precommitted structure—a set of red lines defined *before* the crisis comes. Once inside the engagement, the temptation to rationalize shortcuts or justify moral drift is overwhelming. The only way to resist it is to make the decision in advance.

These boundaries vary by mission, values, and domain. But the question is universal:

*"What actions are we unwilling to take, no matter the gain?"*

Examples include:

- Refusing to deploy disinformation, even if the adversary does.
- Rejecting illicit intelligence or data obtained unethically.
- Declining to partner with known bad actors, even if strategically advantageous.
- Choosing not to retaliate when retaliation would compromise moral clarity.
- Using more force than necessary in a physical conflict.

The point isn't to handicap yourself. It's to protect your *ability to remain who you are.* Because once you've burned that bridge, there is no path back.

## Integrity Under Pressure

The hardest moment to uphold integrity is the one where it feels optional. Especially when your adversary has no such restraint.

This is the terrain of desperation, revenge, existential risk—when the stakes are highest, and the options feel narrow. It's the moment when short-term gain seduces long-term decay.

In adversarial dynamics, you are not only judged by what you achieve, but by *how* you achieve it—and whether your own people can still recognize themselves on the other side.

## Positive Example — Nuremberg: Choosing Law Over Vengeance

In 1945 the Allies faced a brutal decision about how to deal with the men who had directed Nazi crimes. Public appetite for retribution was real, and influential voices floated versions of swift punishment. Others argued that the response had to model the civilized order the war claimed to defend. The trials camp prevailed. The London Agreement of August 8, 1945 and its Charter created the International Military Tribunal (IMT) at Nuremberg, defined the charges— conspiracy to commit crimes against peace, crimes against peace, war crimes, and crimes against humanity—and set procedural guarantees (e.g., notice of charges in a language the accused understands, the right to counsel, and the opportunity to present evidence and cross-examine).

Proceedings opened on November 20, 1945. The next day, U.S. chief prosecutor Robert H. Jackson framed the trial as a test of the *Allies' own character*: the way to answer unprecedented wrongdoing was law, not vengeance. In court, the promise of fairness became practice. Prosecutors built the case chiefly from captured documents, minutes, orders, and ledgers; defendants were represented and cross-examined; and, for the first time at such scale, simultaneous interpretation let four languages share one record in real time. The tribunal took months, not days, because legitimacy requires time.

Verdicts arrived on October 1, 1946. Twelve defendants were sentenced to death, others to prison terms, and three were acquitted —an outcome that showed the court would not criminalize by association. The United Nations affirmed the 'Nuremberg principles' on December 11, 1946, and later tribunals and human-rights instruments drew on the same foundations. The Allies had already won the war; at Nuremberg they won the war of internal integrity. By choosing procedure over revenge, they punished perpetrators without abandoning the norms they claimed to fight for, preserved their own identity under pressure, and converted integrity into operating capital for the postwar order.

Jackson's opening statement for the trials provides a fitting conclusion of the significance of the principle:

> "That four great nations, flushed with victory and stung with injury stay the hand of vengeance and voluntarily submit their captive enemies to the judgment of the law is one of the most significant tributes that Power has ever paid to Reason … We must never forget that the record on which we judge these defendants today is the record on which history will judge us tomorrow."

### Negative Example — The Corporate Culture of Enron

The collapse of Enron is often remembered as a financial failure. In reality, it was a failure of integrity—a system-wide abandonment of ethical boundaries in the pursuit of market dominance.

In the late 1990s, Enron was hailed as an innovative energy company. But inside, its leadership pursued an adversarial strategy not against a single competitor—but against market expectations themselves. They created increasingly complex schemes to hide debt and inflate profits, manipulating not just financial reports but public trust.

These actions weren't isolated. They were *institutionalized.*

Employees were encouraged to beat quarterly targets by any means necessary. Executives sold off stock while assuring investors of stability. The culture became one of strategic deceit—justified, at every level, as necessary to "win."

For a while, it worked. Stock soared. Media praised. Awards followed.

And then, the system collapsed. Once whistleblowers and journalists exposed the fraud, Enron unraveled within weeks. Its auditor, Arthur Andersen, collapsed after its 2002 indictment (the conviction was later overturned in 2005), having already lost most clients and personnel. Thousands of employees lost retirement savings. Executives faced criminal convictions. The company that once seemed unstoppable had, in fact, hollowed itself out.

Enron's failure wasn't just a case of getting caught. It was the predictable result of a system that stopped asking what kind of organization it wanted to be. It had abandoned Principle 28. And by the time it realized what it had lost, it had nothing left to defend.

## Principles in Action — Reflection Questions

1. What actions or tactics would we refuse to use—even if they guaranteed short-term success?
2. Have we clearly articulated our ethical red lines to our team—and would they recognize them under pressure?
3. In what ways does our current operational culture incentivize compromise of values?
4. What would it mean for us to "win" a conflict, but lose our integrity in the process? Would that still be victory?
5. Who in our structure is responsible for upholding our ethical identity—and are they empowered to resist when the pressure comes?

## Sources

- The Economist. 2006. "Enron's Legacy." *The Economist,* January 2006.
- Eichenwald, Kurt. 2005. *Conspiracy of Fools: A True Story.* New York: Broadway Books.
- "Enron Scandal." 2002. *PBS Frontline: The Corporate Scandal Sheet.*
- Gospel of Mark. 1982. *The New King James Version.* Nashville: Thomas Nelson.
- International Military Tribunal. 1947. *Trial of the Major War Criminals before the International Military Tribunal, Nuremberg, 14 November 1945–1 October 1946.* Vol. 1 (Judgment) and Vol. 2 (Jackson's Opening). Nuremberg: IMT.
- McLean, Bethany, and Peter Elkind. 2003. *The Smartest Guys in the Room: The Amazing Rise and Scandalous Fall of Enron.* New York: Portfolio.
- Nietzsche, Friedrich. 1966. *Beyond Good and Evil: Prelude to a Philosophy of the Future.* Translated by Walter Kaufmann. New York: Vintage Books. Originally published 1886.
- United Nations. 1945. "Agreement for the Prosecution and Punishment of the Major War Criminals of the European Axis, and Charter of the International Military Tribunal (London Agreement), 8 August 1945." London: United Nations.
- United States Holocaust Memorial Museum (USHMM). 2020. "International Military Tribunal at Nuremberg." Washington, DC: USHMM.

# PRINCIPLE 29
## FINISH SMARTLY IN VICTORY OR DEFEAT

*"The hardest game to win is a won game."* — Emanuel Lasker
(Attributed)

I n adversarial dynamics, the final moments are often the most treacherous. World Chess Champion Emanuel Lasker captured the most profound part of this challenge with his famous insight, which speaks to a deep truth: the greatest danger in a conflict is often not the adversary, but the complacency that comes with a lead. When the shape of the outcome becomes visible, so do the greatest distortions. Overconfidence and despair are equally dangerous lenses that blur decision-making and distort posture at precisely the moment when clarity matters most. Just as overconfidence can turn victory into defeat, a stubborn refusal to accept a loss can turn a manageable setback into a catastrophe. *Finish Smartly in Victory or Defeat* is the discipline of mastering both sides of this final phase. It is about closing with precision—whether you're about to win or about to lose.

This principle is not about victory. It is about *closure*. It demands that we overcome two great internal enemies: the complacency Lasker warned of as victory seems inevitable, and the stubborn pride that comes with looming loss. Both are forms of emotional inertia. And both can take a manageable outcome and transmute it into something strategically destructive.

Some wins are lost in the final moments. Some defeats are turned into long-term ruin. This principle exists to prevent both.

## The Two Primary Failure Modes

Almost every bad finish falls into one of two camps.

1. **Complacency in Victory.** This is the failure of those who assume the game is over before it is. They have the lead. They feel momentum. They coast. Instead of closing decisively, they shift into "protect mode"—playing not to lose rather than playing to win. This posture invites risk. It gives the adversary room to recover. It stalls initiative. Even worse, it signals weakness at the moment when strength should be most visible.

2. **Stubbornness in Defeat.** This is the failure of those who cannot walk away. They've lost, but they refuse to accept it. Fueled by sunk costs, ego, or fear of humiliation, they keep fighting long after the outcome is clear. This error converts a small loss into a catastrophic one. It burns credibility. It destroys resources. And worst of all, it damages the structure of future recovery.

Both of these errors stem from the same root: the inability to detach from emotion in the endgame. And both are correctable.

## The Discipline of Finishing a Win

When the adversary is off balance, when their cohesion is fractured and their morale is fading, that is not the time to coast. It is the time to apply focused pressure. Winning is not enough. You must *lock in the win*.

The difference between a temporary success and a strategic victory often lies in the final moves. A closing posture must be intentional: decisive, sharp, and measured. When you relax before the engagement is fully secured, you invite resurgence. A flailing adversary given room to breathe will often recover in ways you don't expect.

I once led a team competing for a major federal contract. With a week to go, we had a dominant position: stronger relationships, clearer scope, better pricing. The mood in the office was celebratory. People were coasting. But I pressed for maximum effort—down to the formatting of the final deliverables. The goal was not just to win. It was to win *so clearly* that the competitor would be demoralized and less likely to return in the next round. That final stretch of discipline made the difference. We won. But more importantly, we reshaped the field for the next cycle.

The finish matters. A brilliant campaign can still unravel at the edge of the goal line.

### The Art of the Strategic Withdrawal

On the other side of the equation lies an equally demanding skill: knowing when to step away. To withdraw well is not weakness—it is one of the highest forms of strategic control. To walk away from a losing fight with more of your resources intact, your team beaten but not broken, and your posture preserved is not retreat. It is repositioning.

A smart withdrawal begins with clear-eyed evaluation. What is the cost of continuing? What can still be gained? And what might be lost if we persist? If the answers point toward diminishing returns, continuing becomes not courage—but something else.

There is no shame in stepping back from an unwinnable position. The only shame is in throwing away good energy because pride won't allow you to stop.

## Protecting Your Future Posture

A good finish is also about positioning. What matters most is not how the current conflict ends—but what shape you are in when the next one begins.

A bad win can damage your reputation, exhaust your system, or provoke unnecessary retaliation. A clean defeat can preserve morale, protect credibility, and set the stage for eventual resurgence. In an extended campaign of engagements, your finish becomes your opening for the next round.

What matters is not just who has the upper hand today. What matters is who *still has the ability to play* tomorrow.

## Positive Example — Microsoft Smartly Winds Down the Windows Phone

In the early 2010s, Microsoft was losing the mobile operating system war. Despite billions in investment and the acquisition of Nokia's handset division, its Windows Phone platform was struggling to gain even minimal market share. Apple's iOS and Google's Android had already solidified a duopoly, and Microsoft's late entry into the race was proving unsalvageable.

When Satya Nadella became CEO in 2014, he inherited this failure. Rather than protect sunk costs, he chose to wind the effort down. In 2015 Microsoft recorded a $7.6 billion impairment related to the phone business and cut about 7,800 jobs; the company later ended support for Windows 10 Mobile in December 2019. The strategic move was to exit phone hardware and refocus on cross-platform software and cloud.

But this was not a collapse. It was a *disciplined withdrawal*.

Instead of attempting to beat Apple and Google on their own turf, Microsoft pivoted. It began building best-in-class apps—Office, OneDrive, Outlook—for *its competitors' platforms*. It stopped trying to win the unwinnable fight and redirected its immense resources toward areas where it could still dominate.

Today, Microsoft thrives not because it won the mobile OS war—but because it knew when to walk away and finish smartly. It protected its future posture, preserved its assets, and transformed a tactical defeat into a strategic resurgence.

## Negative Example — Rat Control Failure to Finish Smartly in Urban Areas

In the world of pest control, few adversaries are more adaptive than the urban rat, and the fight against them provides a classic lesson in the failure to finish a win. A common pattern often unfolds in major cities.

A public health department will launch an ambitious initiative to eradicate a major rat colony in a high-density neighborhood. Initial operations are aggressive: poison is deployed and nesting sites are destroyed. Within weeks, the visible rat population drops sharply. Crews report success, and with the immediate crisis seemingly averted, resources are reallocated to other pressing issues. The job, it seems, is done.

But it isn't.

What the extermination team has failed to do is finish the kill. Pockets of the colony survive underground. Early efforts often remove the most visible, trap-prone animals, leaving a residual population that is younger, warier, and increasingly neophobic. Without sustained integrated pest management—sanitation, exclusion, and follow-up treatments—populations rebound quickly due to high reproductive rates and learned avoidance of previously used baits and placements.

Within months, the rat population can return to—or exceed—precampaign levels. The survivors have adapted. They shift foraging times, avoid common toxins, and find new, less accessible spaces. The city now finds itself in a harder, more expensive fight than before.

This is a failure of endgame execution. The city had the advantage. It was winning. But it relaxed too early, mistaking progress for closure. In doing so, it turned a partial victory into a long-term strategic loss by creating an enemy that is not only still alive—but now smarter and more resilient than before.

## Principles in Action — Reflection Questions

1. What are our predefined conditions for both victory and defeat? At what point do we lock in our gains or execute a clean withdrawal?
2. In a commanding position, how do we ensure we don't shift into complacency? What mechanisms trigger our "close decisively" posture?
3. In a deteriorating position, do we have a protocol for evaluating when to stop fighting? How do we distinguish discipline from cowardice?
4. Are any of our current decisions being driven by sunk costs or ego, rather than strategic logic?
5. How will the way we exit this conflict—whether as winners or losers—shape our posture, morale, and credibility for the next one?

## Sources

- Corrigan, Robert M. 2015. *Rodent Control: A Practical Guide for Pest Management Professionals.* Cleveland, OH: GIE Media.
- Foley, Mary Jo. 2018. "Satya Nadella's Biggest Decision: Giving Up on Mobile." *ZDNet,* July 2, 2018. http://www.zdnet.com/article/satya-nadellas-biggest-decision-giving-up-on-mobile/
- Gellerman, Bruce. 2011. "The Rats Are Winning." *Living on Earth,* NPR, March 11, 2011. http://www.loe.org/shows/segments.html?programID=11-P13-00010&segmentID=5
- Grynbaum, Michael M. 2007. "City Unveils New Offensive

in War on Rats." *The New York Times,* July 26, 2007. http://www.nytimes.com/2007/07/26/nyregion/26rats.html

- Lasker, Emanuel. 1947. *Lasker's Manual of Chess.* New York: Dover Publications.
- Le Monde. 2015. "Microsoft accuse la plus importante perte de son histoire." *Le Monde,* July 22, 2015. http://www.lemonde.fr/economie/article/2015/07/22/microsoft-accuse-la-plus-importante-perte-de-son-histoire_4693444_3234.html
- Microsoft Support. 2019. "Windows 10 Mobile End of Support: FAQ." Last updated December 10, 2019. http://support.microsoft.com/en-us/windows/windows-10-mobile-end-of-support-faq-8c6574b4-7229-5dba-7554-18c24a68250f
- Nadella, Satya. 2017. *Hit Refresh: The Quest to Rediscover Microsoft's Soul and Imagine a Better Future for Everyone.* New York: Harper Business.

# PRINCIPLE 30
## REGROUP, LEARN, AND REBUILD

*"The world breaks everyone and afterward many are strong at the broken places."* — Ernest Hemingway

No engagement truly ends when it ends. It echoes.

In adversarial dynamics, the real conclusion of a conflict is not marked by the last blow, the final decision, or the signed agreement. It comes only when the entity that endured the engagement has done the work of understanding what happened —and has turned that understanding into adaptation. *Regroup, Learn, and Rebuild* is the process by which raw experience is converted into refined capability.

An organization or individual that survives but learns nothing has not truly survived. They have simply bought time. What determines long-term viability is not whether you won or lost—but whether you evolved.

This principle is the connective tissue between all others. It completes the loop: the insight gained here is what powers your ability to avoid future conflict (Principle 1), to better understand the adversary, to prepare more intelligently, and to act more decisively when the next challenge comes.

**The Three Phases of Postconflict**

Postconflict recovery is not one event. It is a disciplined sequence of three interlocking phases:

1. **Regroup:** Stabilize the system. Take inventory. Recenter your structure—psychologically, operationally, and structurally.
2. **Learn:** Conduct a forensic, egoless examination of what happened, why it happened, and what it means.
3. **Rebuild:** Translate those insights into structural upgrades: new policies, stronger teams, refined doctrine, tighter tools.

Done correctly, these three phases ensure that no sacrifice was wasted, no loss was pointless, and no win leaves you arrogant. They form the engine of strategic renewal.

**Phase One: Regrouping the System**

The first step after any engagement is stabilization. Before you can learn, you have to stop the bleeding. This includes:

- **Securing the perimeter:** preventing further damage, exploitation, or erosion.
- **Accounting for assets:** assessing what remains—your people, resources, credibility, and political capital.
- **Restoring a baseline:** reestablishing communication and psychological orientation so your system can function again.

But regrouping is not only logistical. It is psychological. After a victory, the threat is arrogance—believing you have nothing left to learn. After a loss, it is shame, fear, and fragmentation. Both distort your ability to analyze clearly.

A leader's job in this phase is to stabilize the team *and* the narrative: to reframe what just happened in a way that invites honesty, reflection, and readiness for analysis.

## Phase Two: The Discipline of Honest Learning

This is the heart of the principle. And it is the part that almost everyone gets wrong. The purpose of this phase is not to assign guilt. It is to extract cause.

- Ask not just "what failed?" but "*why* did it fail?"
- Push past the proximate cause—the triggering event—and find the root cause: the culture, assumptions, or incentives that allowed it.
- Treat the data as sacred. Treat the egos as noise.

This kind of learning requires humility, curiosity, and a culture that values truth more than comfort. Organizations that cannot develop these traits will never adapt fast enough to outpace their adversaries.

## Phase Three: Rebuilding, Not Just Repairing

Learning without action is philosophy. Learning *with* action is transformation.

But not all action is meaningful. Most organizations don't rebuild. They repair. They replace the person who failed. They patch the system that broke. They publish a memo. And then, when the same failure emerges six months later in a different form, they act surprised.

Rebuilding requires going deeper. It asks:

- Was the system designed wrong?
- Was the mission unclear?
- Did our incentives encourage this failure?

- What needs to be permanently different for this to never happen again?

And then it institutionalizes the answer. Real rebuilding means changing hiring protocols, rewriting doctrine, reshaping training, or eliminating entire layers of a structure if they no longer serve. It is costly. It is hard. But it is the only way to truly close the loop.

## The Adversarial Feedback Loop

This final principle is what ensures that your system does not stay frozen in one era, one war, one engagement.

The moment you finish an engagement—win or lose—you begin shaping the next one. And how well you regroup, how deeply you learn, and how honestly you rebuild determines whether you evolve … or repeat.

If you do this well, the adversary will never face the same version of you twice. That is evolution. That is strategic maturity.

## Positive Example — Team Rubicon: Veterans Rebuilding Purpose After War

For many veterans, the fight that follows combat is intrapersonal: the loss of structure, purpose, and belonging that held life together downrange. Team Rubicon began in 2010 as a direct response to that aftershock. Team Rubicon is a U.S.-based nonprofit that mobilizes military veterans, first responders, and civilian volunteers—its "Greyshirts"—to deliver rapid disaster relief. But the founders did not set out first to fix disaster relief; they set out to keep *themselves* intact.

**Regroup.** They pulled veterans back into a unit-like container: a clear identity, a shared mission vocabulary, and predictable routines that felt like muster instead of drift. The point of early meetups, onboarding, and role clarity was unification and stabilization—replacing isolation and idle time with community, cadence, and accountability.

**Learn.** Through candid peer conversations and repeated check-ins, they named what the postwar gap really was. It wasn't just missing paychecks; it was missing *mission, tribe,* and *tempo.* They learned that service to others quieted hypervigilance and aimlessness; that clear roles and simple, repeatable procedures restored agency; that standards and shared hardship restitched identity faster than solitary "time off." The lesson was about the enemy within—drift and disconnection. With this information understood, they could then act.

**Rebuild.** They built a vehicle around those insights: a membership and training pathway that turns veterans' skills into steady service; uniforms and symbols that anchor identity; team norms, after-action circles, and leadership lanes that keep purpose durable between deployments. Disaster missions became the arena, not the point. The strategic win was personal and collective integrity rebuilt —veterans finishing one war, facing the intrapersonal aftermath with honesty, and returning to the field as whole people who can serve again tomorrow.

### Negative Example — Mike Tyson After Buster Douglas: Failing to Regroup, Learn, and Rebuild

If *Tyson vs. Douglas* seems a familiar topic, it is by design. We covered the actual fight in Principle 3. That sets the stage nicely for examining the aftermath of that loss. Entering that 1990 bout, Tyson was undefeated in 37 fights; the loss shattered the aura that had carried him through the 1980s. Tyson was arguably still in his prime. He could have used the loss as a wakeup call. He could have focused and salvaged his career trajectory. Tyson did not apply *Principle 30* in a complete and meaningful way. What unfolded was sadly predictable.

**Did not regroup.** Instead of stabilizing his camp and recentering his preparation after the loss to Douglas, Tyson's post-Tokyo period was marked by churn and erratic discipline—patterns he later acknowledged in his memoir (drugs, poor conditioning, unfocused camps). The absence of a settled technical team and routine meant no coherent reset after the shock.

**Did not learn.** The tactical slippage that showed up in Tokyo—less head movement, a thinner jab, reliance on intimidation—was not systematically corrected. Tyson describes repeatedly reverting to shortcuts rather than rebuilding the habits that had once made him hard to hit and easy to set up combinations for. There was plenty to learn from. If he did learn the lessons, he did not apply them.

**Did not rebuild.** Instead of constructing a durable training architecture and governance around himself (coaches, conditioning, accountability), the structure stayed fragile. Flashpoints of poor control bled into outcomes—e.g., the 1999 Orlin Norris bout was ruled a no contest after an after-the-bell foul—signaling process breakdown rather than process repair.

The rest of the 1990s–2000s brought multiple high-profile defeats (including to Evander Holyfield, Lennox Lewis, and late-career losses), underscoring the core lesson of this principle: if you don't regroup, learn, and rebuild after a setback, the next adversaries won't be facing a better version of you—just a more vulnerable one.

### Principles in Action — Reflection Questions

1. After a major engagement, what is our structured process for regrouping—both logistically and psychologically?
2. How honest are our after-action reviews? Do they seek truth or protect egos?
3. Do we focus on root causes or settle for proximate ones? Are we willing to change fundamental parts of our system if needed?
4. When we rebuild, are we repairing the past or designing for the future? What's the difference?
5. How do we institutionalize the lessons we learn so they persist beyond individual memory or leadership turnover?

## Sources

- BoxRec. n.d. "Mike Tyson." http://boxrec.com/en/proboxer/474
- Eastridge, Brian J., Donald Jenkins, Stephen Flaherty, Henry Schiller, and John B. Holcomb. 2006. "Trauma System Development in a Theater of War: Experiences from Operation Iraqi Freedom and Operation Enduring Freedom." *Journal of Trauma: Injury, Infection, and Critical Care* 61 (6): 1366–72. https://doi.org/10.1097/01.ta.0000245894.78941.90
- Freeman, Mike. 1999. "In Bizarre Ending, Tyson Bout Is Declared a No Contest." *The New York Times,* October 24, 1999. http://www.nytimes.com/1999/10/24/sports/in-bizarre-ending-tyson-bout-is-declared-a-no-contest.html
- Hemingway, Ernest. 1929. *A Farewell to Arms.* New York: Charles Scribner's Sons.
- Sports Illustrated. 2015. "Tyson vs. Douglas: 25th Anniversary." *Sports Illustrated,* February 10, 2015. http://www.si.com/boxing/2015/02/10/mike-tyson-buster-douglas-25th-anniversary
- Team Rubicon. n.d. "About Us." http://teamrubiconusa.org
- Tyson, Mike, and Larry Sloman. 2013. *Undisputed Truth.* New York: Blue Rider Press.

# CASE STUDIES
## UTAD'S FINAL EXAM

*"It doesn't matter how beautiful your guess is, it doesn't matter how smart you are. If it doesn't agree with experiment, it's wrong."* —Richard P. Feynman

Theory is only as good as the trials it survives. Everything in this book—every principle, every axis, every structural phase—has been designed to help make sense of adversarial dynamics across domains. But no model earns its keep without full contact. This chapter marks that contact.

What follows are three case studies drawn from radically different worlds: insurgent revolution, cyberwarfare, and global disease eradication. They were not chosen for drama, politics, or elegance. They were chosen because each represents the full structural arc of conflict. In each case, the UTAD framework doesn't merely apply—it illuminates. Across all three, you will find the thirty principles in action. Each example here is grounded in public fact and supported by reputable, verifiable scholarship.

You'll notice these are narrative reconstructions. They are focused, condensed, and intentionally incomplete—not because detail doesn't matter, but because too much of it can bury the signal.

The first case takes us to the American Revolution—a long-form asymmetric campaign in which coordination, internal discipline, and moral framing outweighed numerical strength. Through it, we see how strategic clarity, alliance-building, environment-shaping, and adaptive resistance coalesced into a national transformation.

The second case enters the covert terrain of Stuxnet, the landmark U.S.-Israeli cyber operation that quietly sabotaged Iran's nuclear centrifuge infrastructure. Here, the principles of deception, timing, resource denial, and precision control disruption are not just applied—they define the entire engagement.

The third case pulls back into the long game: the global effort to eradicate poliovirus. This is no metaphor. It is a conflict fought against a non-volitional adversary—one that adapts through biology, not strategy—and yet the dynamics remain adversarial. What emerges is a generational campaign grounded in strategic stamina, coordination, learning, and the unglamorous but essential discipline of finishing what you start.

Across these cases, you will see how conflict follows similar patterns. Not just in war rooms or crisis chambers, but in code, in cities, in immune systems. And as you read, you will hopefully begin to see the deeper truth: that while the tools change and the context shifts, the underlying principles—the laws of adversarial structure—do not.

That is the power of a unified theory.

～

## Case Study #1: The American Revolution

In early 1775, the thirteen colonies on the eastern edge of North America looked like a fragile fringe of Britain's empire—scattered, under-resourced, and politically divided. At that point they had no standing national army, no navy, and no central treasury. The Second Continental Congress moved quickly: it established the Continental Army on June 14, 1775, appointed joint treasurers on July 29, 1775, and on October 13, 1775, Congress ordered two vessels to be armed; broader naval authorization followed later that year. Britain, by contrast, was one of the most dominant great powers of the age, with a global fleet, hardened redcoats, and centuries of war-fighting experience. To rebel against such a force looked suicidal.

Yet by 1783 the colonies had secured independence. Their victory was far from inevitable. It was achieved through adaptation, coalition-building, discipline, and a structure of thinking that reflected nearly every principle of adversarial dynamics.

## I. Before the War: Shaping the Conditions

The Revolution did not begin with a musket volley. For more than a decade, colonial leaders tried to *prevent conflict proactively (Principle 1)*. They petitioned Parliament, convened congresses, and organized Committees of Correspondence to resolve grievances without bloodshed. This restraint was strategic: once violence began, it would transform everything.

When reconciliation collapsed, the colonies moved to *define victory with precision (Principle 2)*. The Declaration of Independence in July 1776 did not merely reject British authority; it declared the colonies to be "Free and Independent States," defining independence as the end state. That clarity became a compass, anchoring subsequent decisions.

They also worked to *shape the environment to their advantage (Principle 11)*. Parallel legislatures, independent militias, and circulating pamphlets reframed legitimacy. By 1775–76 Britain was no longer suppressing disorder—it was facing an alternative political structure.

To secure support, revolutionaries had to *engage stakeholders with intentionality (Principle 10)*. Paine's Common Sense and town meetings were not casual persuasion; they were designed to flip neutral colonists into committed allies.

Just as critical was the need to *align internally (Principle 5)*. Thirteen separate colonies, each with divergent economies and fears, had to hold together. Washington's skill lay not only in strategy but in sustaining an army, a Congress, and a fragile unity that might otherwise have splintered.

Finally, they pursued external backing. Franklin's diplomacy in Paris, alongside Vergennes and French court politics, allowed the colonies to *ally with purpose (Principle 6)*. France's primary motivation was strategic rather than ideological; both powers shared Britain as the adversary. The alliance rested on complementary strengths and proved decisive.

## II. Becoming a Force Worth Fighting

Once war was unavoidable, survival required capability. At Valley Forge in 1777–78, Friedrich von Steuben introduced standardized drills; in 1779 he codified them in Regulations for the Order and Discipline of the Troops of the United States (the "Blue Book"). This was an effort that aspired to *build capability that renders strategy irrelevant (Principle 3)*—turning raw recruits into a disciplined force able to execute.

Washington also grasped the value of intelligence. Through the Culper Ring and other networks, he deployed codes, dead drops, and invisible ink to *know the adversary better than they know you (Principle 4)*. Against a stronger enemy, foresight and deception were force multipliers.

The revolution also revealed the risk of betrayal. The case of Benedict Arnold nearly cost West Point. Americans had to *account for insider risk (Principle 9)* with layered trust and safeguards.

Planning extended beyond the battlefield. Congress modeled contingencies for famine, invasion, and leadership collapse—an example of *plan for all scenarios (Principle 7)*.

But theory is only as good as the trials it survives. The raw colonial militias faced their first major trial at the Battle of Bunker Hill in June 1775, which provided an opportunity to test against *realistic resistance (Principle 8)*. The engagement was a brutal, live-fire test against elite British regulars. Though the Americans technically lost the ground, the test was a profound strategic success. It was not fatal. It provided a crucial psychological victory, validating their defensive tactics. It exposed a critical weakness in their supply chain when they ran out of ammunition—an important lesson to be carried into future battles. The staggering casualties also served as a harsh lesson for the British, prompting greater caution in frontal assaults—especially around Boston.

### III. Gaining Momentum: Maneuver and Deception

With structure in place, the Americans began to press outward. They learned to *exploit the environment (Principle 15)*—fighting from forests, ambushing supply trains, and using winter weather as a weapon.

Washington mastered timing. His Christmas night crossing of the Delaware in 1776 was a case of *time your actions strategically (Principle 16)*: striking when Hessian forces at Trenton were vulnerable from fatigue and complacency.

The Americans embraced initiative: *strike first and sustain pressure (Principle 12)* at Trenton and Princeton, forcing the British onto the defensive. They consistently *targeted weakness and avoided strength (Principle 13)*, hitting isolated garrisons rather than seeking glory in open battles.

Operational security demanded they *reduce their exposure (Principle 14)*—frequent relocation of commanders, dispersed supply caches, and compartmentalized communications denied the British any decisive strike.

Over time, their ability to *adapt quickly to changing conditions (Principle 17)* became evident. The force that lost New York in 1776 was not the one that marched to Yorktown in 1781.

Deception crowned these efforts. Before the Yorktown campaign, Washington staged preparations for a strike on New York while secretly marching south—a textbook use of *deceive to force mistakes (Principle 19)*. The British concentrated on the wrong front.

When a superior British force made Fort Ticonderoga indefensible in 1777, the American commander chose to *sacrifice tactically to gain strategically (Principle 20)* by abandoning the artillery, supplies, and symbolic fortress itself in order to save his about 2,500–3,000 Continentals (plus militia) from certain destruction. Though the retreat was a blow to public morale, those preserved forces reinforced the American army that later defeated the British at the Battles of Saratoga—the victory that secured French intervention and turned the tide of the war.

Concurrently, the patriots worked to *stretch the adversary's defenses (Principle 18)*. France's entry in 1778 forced Britain to divide attention among the Channel, the Caribbean, and coastal North America. In the South, Nathanael Greene's campaign—marching, feinting, and refusing decisive battle—pulled British columns across hundreds of miles, while militia raids harassed supply lines. The net effect was dispersion: ships, regiments, and senior focus spread thin enough that Cornwallis could be isolated and fixed at Yorktown.

## IV. Endgame: Pressure, Collapse, and a Clean Finish

By late 1781, the Americans focused on attrition. They aimed to *drain the adversary's resources systematically (Principle 21)*. Britain could still win battles, but each was bleeding money, manpower, and political will at disproportionate rates.

At Yorktown, they *disrupted the adversary's control center (Principle 22)*. With French naval dominance in the Chesapeake cutting off relief and supply, Cornwallis was isolated from higher command. The British body remained, but its head was silenced.

The final collapse was psychological. The revolutionaries worked to *break the adversary's will to fight (Principle 23)*. After Yorktown, the North ministry fell; on February 27, 1782, the House of Commons passed a non-binding motion urging cessation of offensive operations in America. By 1783 the Treaty of Paris ended the war—not because Britain was destroyed, but because it judged victory no longer worth the cost.

In March 1783, as anonymous circulars urged officers toward mutiny over arrears, Washington defused the Newburgh crisis by appealing to honor and shared sacrifice—famously remarking that he had "not only grown gray but almost blind in the service of my country." By choosing to *uphold integrity (Principle 28)* at the decisive moment, he shut down a potential military coup and preserved the fragile legitimacy of the new republic.

Washington showed restraint. He chose to *finish smartly in victory or defeat (Principle 29)*. He avoided overreach, consolidated gains, and secured a decisive settlement rather than pressing for more.

Throughout this phase, American leaders continued to *calculate risk and reward (Principle 24)* in every action, resisting temptations to overexploit British weakness. Washington personally *maintained composure under pressure (Principle 25)*, setting the emotional tone for both soldiers and civilians.

He also worked to *nurture critical resources (Principle 26)*—keeping an underpaid, underfed, exhausted army intact through personal attention and care. Campaigns were *paced for sustainability (Principle 27)*; instead of burning out in pursuit of glory, the revolutionaries fought in measured arcs that could be sustained over years.

## V. After the War

With independence won, the greater test remained: to build a durable polity. The first framework, the Articles of Confederation, soon proved inadequate. The Constitutional Convention of 1787 was an act to *regroup, learn, and rebuild (Principle 30)*, replacing the Articles with a stronger constitution that limited executive power and protected liberties under law.

## Case Study Summary:

The American Revolution demonstrates how a weaker actor, guided by discipline and structure, could defeat a superior adversary. Every UTAD principle neatly finds expression here. The colonies did not prevail by overwhelming force. They prevailed by thinking struc-turally, sustaining cohesion, and finishing cleanly when the moment arrived.

~

## Case Study #2: Operation Stuxnet

Iran built its Natanz enrichment plant to be untouchable: a fortress of earth and concrete shielding thousands of centrifuges spinning uranium hexafluoride. Layers of guards, compartmentalized access, and a digital air gap separated the facility and its computers from the outside world. To most observers, it was unhackable. What followed was the *first widely reported and confirmed* digital weapon to cause physical destruction inside an industrial system.

## I. Preventing the Fight

Before code was written, the United States and its partners attempted to *prevent conflict proactively (Principle 1)*. Between 2003 and 2008, negotiators offered sanctions relief in exchange for suspending enrichment, and monitored suspension agreements were explored. When these efforts failed, Washington and Jerusalem turned to a quieter option: stealth sabotage rather than overt escalation. The operation began by *defining victory with precision (Principle 2)*. The goal was not to crash networks, steal blueprints, or signal power. It was narrower: to delay or derail Iran's nuclear program by silently degrading centrifuges from within, while avoiding attribution.

## II. Building a Weapon That Did Not Exist

The scale of the task demanded capability, not improvisation. Planners set out to *build capability that renders strategy irrelevant (Principle 3)*. They reverse-engineered Siemens STEP 7 controllers used inside the facility, constructed mockups of centrifuge cascades, and tested payloads against real rotor physics. By the time the weapon launched, it was a functioning industrial sabotage system. This was not the work of one country. It was *widely attributed* to a joint U.S.–Israel operation (neither government has formally acknowledged this) to *ally with purpose (Principle 6)*. The Americans brought zero-day exploits, cyber tradecraft, and strategic targeting. The Israelis contributed regional intelligence, operational access, and cultural fluency. Each partner covered the other's blind spots. Planners also *shaped the environment to their advantage (Principle 11)*. They studied the exact control hardware used at Natanz and tailored code for those systems, validating it against representative Siemens controllers. By the time Stuxnet arrived, the terrain was already prepared.

The weapon was designed to *exploit the environment (Principle 15)* by turning Natanz's own systems into weapons. It leveraged the specific physics of the IR-1 centrifuges, subtly changing rotor speeds to induce harmonic vibrations that would destroy them over time. Further, it exploited the procedural environment of the air-gapped facility, correctly predicting that USB drives used by outside contractors would *likely* serve as the physical bridge to the isolated network. They deliberately *targeted weakness and avoided strength (Principle 13)*. No attempt was made to breach the fortified perimeter or overpower security forces. The exploitable surface lay in the programmable logic controllers—the invisible point where digital instructions met fragile centrifuge physics.

The operation's success depended on more than just technology; it required *engaging stakeholders with intentionality (Principle 10)* by building a fragile coalition between rival agencies within both the U.S. (like the NSA and CIA) and Israel (like Mossad and Unit 8200). Securing the combined cyber, nuclear, and human intelligence expertise from these fiercely independent organizations—and getting them to commit their best resources to a single, high-risk covert action—was a critical victory in itself.

Stuxnet was engineered to make its damage appear as normal industrial wear, operator error, or faulty equipment. This manipulation ensured that observers would not raise an alarm about a cyberattack, thereby protecting the operation's secrecy. The campaign relied on deep insight: to *know the adversary better than they know you (Principle 4)*. Stuxnet not only sabotaged centrifuges but it also replayed fake telemetry to plant operators, showing equilibrium while rotors tore themselves apart. This was knowledge of both machine and human response. Security also demanded they *account for insider risk (Principle 9)*. Delivery most likely depended on technicians and contractors carrying removable media. To guard against leaks or compromise, access was compartmentalized and insiders were screened.

## III. Engineering for War

The weapon was not a crude strike but a system designed with discipline. Its authors worked to *calculate risk and reward in every action (Principle 24)*. They burned *reportedly four* zero-day vulnerabilities— expensive assets that could have fueled other campaigns. That risk was accepted because the payoff was judged decisive. Commanders overseeing the project had to *maintain composure under pressure (Principle 25)*. Diplomatic tensions spiked as development dragged on. Internal debates tested patience. Still, the program moved forward methodically. To protect the mission, operators took steps to *reduce their exposure (Principle 14)* by eliminating any need for local presence at the site. The code would do all the work without risking American or Israeli lives or equipment.

The code was built with a kill date (June 24, 2012), ensuring it would deactivate after a set period to prevent indefinite spread. Knowledge of the program was highly compartmentalized between different agencies and even between the U.S. and Israeli partners to minimize the impact of a potential leak. This secrecy was a defensive measure to make the operation itself a difficult target to identify and attribute. The design also reflected a *plan for all scenarios (Principle 7)*. What if the worm was discovered early? What if it missed its target? The code included fallback conditions and safeguards. Before deployment, it was tested against *realistic resistance (Principle 8)*. Stuxnet's creators did not rely on simulation alone—they reportedly ran the worm against live centrifuge arrays in Israel, observing real damage before unleashing it in Iran.

## IV. The Deployment

Air gaps are never perfect. At Natanz, the worm *likely* entered via infected USB drives, possibly through contractors, bridging the air gap. Once inside, it spread discreetly across internal systems. The coalition had to *align internally (Principle 5)*. Two governments, multiple agencies, and competing timelines had to act as one. Coordination extended down to drop windows and synchronized payload triggers. Resources were limited. Zero-days are finite tools. The attackers deliberately *nurtured critical resources (Principle 26)*, stretching the lifespan of the exploits by metering deployments and tuning the payload to minimize detection.

The worm itself was designed to *time your actions strategically (Principle 16)*. It waited silently until centrifuges reached specific speeds. Then it triggered bursts of sabotage at just the right moment to maximize mechanical stress. It was also built to *deceive to force mistakes (Principle 19)*. While centrifuges spun erratically, the operators' screens displayed stable readings. Each misinformed adjustment deepened the damage. The attack did not happen once but in waves. Stuxnet was engineered to *strike first and sustain pressure (Principle 12)*, embedding itself to reinitiate sabotage over months. The ultimate effect was to *drain the adversary's resources systematically (Principle 21)*. Each destroyed centrifuge had to be replaced at significant cost in time, money, and materials. More importantly, the unexplained failures drained cognitive resources, forcing Iran's best engineers into a frustrating cycle of diagnosing phantom mechanical faults and questioning their own suppliers, diverting them from the primary goal of enrichment.

It also had to *pace actions for sustainability (Principle 27)*. A catastrophic crash might have revealed the intrusion immediately. Instead, the worm produced gradual failures that looked like ordinary wear and tear. Under stress, Iran was forced to *stretch its defenses (Principle 18)*. Engineers searched for faulty equipment. Security services hunted spies. Resources scattered across misdiagnosed problems. The real cause—malware inside controllers—remained hidden.

All the while, Stuxnet *disrupted the adversary's control center (Principle 22)*—the computer systems themselves that ran the entire facility. Iranian decision-makers saw clean telemetry even as physical machines failed. Their command network was blinded. By making Iran's own advanced technology betray its creators, the operation was engineered to *break the adversary's will to fight (Principle 23)*. The psychological goal was to sow doubt, paranoia, and a sense of technological futility. By making the nuclear program appear unreliable and cursed by mysterious technical failures, the attackers aimed to erode the confidence and morale of Iran's scientific and political leadership, making them question if the program was worth the immense and frustrating effort.

The code was even designed to *adapt quickly to changing conditions (Principle 17)*, altering behavior if it encountered unexpected system states. The choice to use four zero-days showed a willingness to *sacrifice tactically to gain strategically (Principle 20)*. Once exposed, those exploits could never be reused. But the payoff—crippling a hardened nuclear facility without a shot fired—was worth the cost.

## V. Closing the Loop—and Missing It

Operationally, the sabotage succeeded. About 1,000 centrifuges were taken offline or replaced. Iran's enrichment program was *delayed by months (estimates vary)*. But Stuxnet's propagation behavior allowed it to escape Natanz. It was discovered in June 2010 by the Belarusian firm VirusBlokAda, dissected, and publicized. Here, the attackers failed to *finish smartly in victory or defeat (Principle 29)*. Although the malware included a kill date (June 24, 2012), secrecy—the center of gravity—was lost once the worm propagated beyond its target.

## VI. Aftermath and Memory

The consequences extended far beyond Natanz. Iran launched its own cyber program. The United States formalized offensive cyber doctrine. Cyber Command institutionalized lessons. Each actor had to *regroup, learn, and rebuild (Principle 30)*. For those who built it, Stuxnet became a model of preemptive, limited action. Its designers argued they had *upheld their integrity (Principle 28)* by keeping the mission proportional and narrowly scoped. Yet critics called it reckless: the opening of a dangerous new frontier, where sabotage blurred into war. What began as code on a USB stick became a campaign that crossed air gaps, paralyzed decision-making, and altered the tempo of global conflict. It was not just an intrusion. It was doctrine.

### Case Study Summary:

Operation Stuxnet demonstrated that adversarial dynamics do not vanish in cyberspace—they intensify. The campaign relied on precision, patience, alliance management, and structural thinking. Every UTAD principle appeared. The operation succeeded tactically but faltered in its finish, showing how even the most advanced design can fail if closure is mishandled. Yet it also shows the stability of UTAD's strategic framework: if most principles are executed well, some failures can be sustained.

~

### Case Study #3: The Global Fight to Eradicate Polio

As a member of Rotary International, a key contributor in the fight against polio, this case study feels personal for me. Poliomyelitis is an ancient adversary. It spreads through contaminated water and food, often moving silently through crowded communities. Most infections cause no symptoms. About 1 in 200 infections leads to irreversible paralysis; among those paralytic cases, 5–10% result in death when breathing muscles are immobilized. There is no cure—only prevention.

For much of history, polio was simply a fact of life. That changed in the late twentieth century when a coalition of states, institutions, and volunteers attempted something once thought impossible: eradication. This campaign was not just medical—it was structural, requiring long-term endurance, coordination across over 200 countries and territories, and the disciplined application of principles recognizable in any form of conflict.

## I. Launching the Campaign

In 1985 Rotary International launched PolioPlus, the first global private initiative to vaccinate children against polio. In 1988, the World Health Assembly (Resolution 41.28) created the Global Polio Eradication Initiative (GPEI), bringing together WHO, UNICEF, CDC, and Rotary, later joined by the Gates Foundation in 2007 and Gavi in 2019. From the beginning, the alliance worked to *define victory with precision (Principle 2)*. The target was not containment but eradication—reducing cases worldwide to zero.

The effort also sought to *prevent conflict proactively (Principle 1)* by immunizing the 99 percent of the world already free of polio, closing off terrain where the virus might re-emerge.

Its coalition was built to *ally with purpose (Principle 6)*. Each partner contributed a distinct strength: Rotary mobilized funding and volunteers; UNICEF leveraged community trust; the CDC provided epidemiology; WHO coordinated globally. To hold this coalition together, GPEI leaders worked constantly to *align internally (Principle 5)*, producing joint action plans, standard operating procedures, and shared monitoring frameworks.

The campaign had to *know the adversary better than they know you (Principle 4)*. A worldwide surveillance system was built to detect not only clinical paralysis but also traces of poliovirus in sewage, often weeks before symptoms appeared. That intelligence allowed interventions before outbreaks took root.

Capacity was also central. Millions of vaccinators, logisticians, and translators were trained to *build capability that renders strategy irrelevant (Principle 3)*. With such reach, the virus found fewer places to hide.

## II. Testing Systems, Anticipating Breakdowns

The campaign faced threats not only from the virus but also from within. Misinformation, corruption, or worker fatigue sometimes undermined delivery. Leaders had to *account for insider risk (Principle 9)* by instituting monitoring, audits, and accountability mechanisms.

They also ran drills to *test against realistic resistance (Principle 8)*. National Immunization Day simulations exposed weaknesses in logistics chains, reporting systems, and cold storage. These rehearsals allowed adjustments before real emergencies.

Contingency planning was constant. Civil war, vaccine refusal, refrigeration failure—all had to be anticipated. This was an example of *plan for all scenarios (Principle 7)*, ensuring that no single breakdown collapsed the whole effort.

## III. Sustaining Pressure

To win a decades-long campaign, resources had to be carefully stewarded. Cold chains were expanded redundantly, donor networks refreshed, and volunteer rotations scheduled to *nurture critical resources (Principle 26)*. Leaders deliberately *paced actions for sustainability (Principle 27)*, avoiding burnout in a struggle that stretched across generations.

The coalition also worked to *shape the environment to their advantage (Principle 11)* through education campaigns and goodwill ambassadors, paving the way for acceptance of vaccinations in otherwise-resistant areas.

When transmission persisted, the GPEI chose to *sustain pressure (Principle 12)*. National Immunization Days mobilized millions of health workers to vaccinate every child under five across entire countries in just days, overwhelming the virus before it could regroup.

Operations *targeted weakness and avoided strength (Principle 13)*. Resources were concentrated on vulnerable zones—urban slums, border crossings, and nomadic routes—where small pushes could break transmission.

They also sought to *reduce exposure (Principle 14)*. Mobile teams worked discreetly, and "mop-up" campaigns encircled outbreak sites quickly to limit spread.

Timing was critical. Campaigns paused during floods or fighting, then resumed when conditions shifted. This was *time your actions strategically (Principle 16)*—choosing windows where success was possible.

In resistant regions like northern Nigeria and Pakistan, the campaign confronted organized opposition fueled by rumor and politics. To win, the GPEI applied a dual strategy. First, it worked to *engage stakeholders with intentionality (Principle 10)*, recruiting respected Islamic scholars who issued fatwas declaring vaccination religiously compliant and partnering with local influencers to build community trust from the inside out. Simultaneously, these same actions served to *disrupt the adversary's control center (Principle 22)*; by co-opting the key nodes of the resistance movement, the campaign fractured anti-vaccine narratives and neutralized the command system that spread misinformation, turning the opposition's greatest strength—its leaders—into a vector for its defeat.

The campaign also had to *stretch the adversary's defenses (Principle 18)* by fighting on multiple fronts, addressing the virus, removing contamination sources, and confronting the virus' human enabler.

As coverage rose, operations focused on *draining the adversary's resources systematically (Principle 21)*. Each new layer of immunity shrank the pool of susceptible hosts, starving the virus of opportunity.

## IV. Under Fire

Polio eradication was not without violence. In Pakistan, Nigeria, and Afghanistan, vaccinators were sometimes attacked and killed. Here leadership had to *maintain composure under pressure (Principle 25)*— pausing campaigns, negotiating with local leaders, then re-engaging when conditions allowed.

While a virus has no consciousness, its human enablers do. The campaign's final push worked to *break the adversary's will to fight (Principle 23)* by dismantling the social and political support for vaccine refusal. Relentless information campaigns, backed by local religious and community leaders, socially isolated anti-vaccine proponents. As case numbers plummeted and the benefits of vaccination became undeniable, the will of communities to resist crumbled. Victory was achieved not just by defeating the virus in children's bodies, but by defeating the ideas that protected it in adults' minds.

Sometimes it was necessary to *sacrifice tactically to gain strategically (Principle 20)*. Local operations were suspended to preserve community trust or save lives, ensuring the broader campaign could continue.

Every decision required leaders to *calculate risk and reward in every action (Principle 24)*—weighing the safety of workers, the fragility of local trust, and the epidemiological stakes.

The program also *adapted quickly to changing conditions (Principle 17)*. From shifting to monovalent OPVs to introducing bivalent vaccines in 2016, the GPEI repeatedly adjusted to new realities.

It also learned to *exploit the environment (Principle 15)*, positioning vaccinators at bus stations, border crossings, and markets—turning ordinary movement into surveillance and opportunity.

Even vaccine design reflected adaptation. To combat the dominant wild strain, health authorities shifted formulations. At times, communication strategy leveraged opponents' assumptions—using selective disclosure and message timing that led organizers of resistance to misread momentum—*deceive to force mistakes (Principle 19)*— while never falsifying facts or breaching ethical lines.

## V. Closing the Fight

By the 2010s, the map of wild polio had contracted to just a handful of regions. India, once considered the hardest front, recorded its last case in January 2011 and was certified polio-free on March 27, 2014 by WHO after three years of surveillance—a disciplined example of *finish smartly in victory or defeat (Principle 29).*

Throughout, the campaign worked to *uphold its integrity (Principle 28).* Even under pressure, leaders refused to cut ethical corners, building long-term trust essential for success.

Every setback became a lesson. Outbreaks were analyzed, tactics revised, and policies updated. This continuous cycle reflected *regroup, learn, and rebuild (Principle 30).*

As of 2025, wild poliovirus type 1 remains endemic only in Afghanistan and Pakistan. Global case numbers have been driven to the edge of extinction. The campaign has not only brought the virus to the edge of extinction but has left behind a logistical, organizational, and bio-surveillance architecture ready for future fights.

### Case Study Summary:

Polio has no mind, no will, no strategy. Yet remarkably, the full scope of the thirty principles proved essential in the fight against it. Against the virus itself, operational endurance, adaptability, and preparation proved decisive. Against human resistance—misinformation, fear, and violence—the campaign confronted an adversary every bit as intentional as any army. The lesson is clear: even when the enemy does not think, strategic structure still matters. Victory was not inevitable. It was earned by treating a virus as an adversary that could be maneuvered against, starved, and finally cornered until it had nowhere left to go.

～

THREE DOMAINS. THREE ADVERSARIAL CAMPAIGNS. ONE COMMON thread.

None of these outcomes—defeating polio, delaying Iran's nuclear capability, or securing American independence—were inevitable. Each involved long odds, real risks, and adversaries that adapted over time. Victory in these cases wasn't guaranteed by wealth, or morality, or even superiority in force. It was earned through structure—through a pattern of disciplined behavior under pressure.

The global effort to eradicate poliovirus didn't succeed because the world simply cared enough. It succeeded because stakeholders aligned, logistics scaled, and trust was earned village by village. Stuxnet didn't reshape cyber conflict because it was clever—it did so because it fused precision, patience, and domain fluency into a single, coherent strategy. The American Revolution didn't prevail because it had the stronger army—it prevailed because it stretched British defenses, fractured their will, and built internal coherence faster than the crown could react.

None of these campaigns used all thirty UTAD principles perfectly. Some were violated. Some were underdeveloped. But the patterns are unmistakable: the most decisive moments in each campaign can be understood through this lens. In fact, it is difficult to find a major strategic success—or failure—in these stories that falls outside the UTAD framework. Every turning point involved a shift in posture, a breakdown in timing, a collapse in trust, a misread environment, or a missed recovery cycle.

Each case spanned the full arc of adversarial dynamics: from preengagement posturing to live adaptation to postconflict regeneration. And in each, the principles proved legible. Not theoretical. Not speculative. Operational.

So the challenge now passes to the reader: Look again. In these case studies—or others. If there are meaningful, repeated principles of adversarial success that fall outside this framework, then UTAD has room to grow. But if not—if every structural advantage, every decisive action, every critical collapse can be traced back to one of these thirty principles—then UTAD is more than a theory. It is a framework worth codifying.

## Sources

- Abraham, Thomas. 2018. *Polio: The Odyssey of Eradication.* New York: Columbia University Press.
- Adams Papers Digital Edition (Massachusetts Historical Society). 1782/2018. *Correspondence Referencing the House of Commons Debate and Motion of 27 February 1782 to Discontinue Offensive War in America.* Boston: MHS.
- Aylward, Bruce. 2012. Interview by Rotary International. Video, May 2012.
- Bailyn, Bernard. 1967. *The Ideological Origins of the American Revolution.* Cambridge, MA: Harvard University Press.
- Bergman, Ronen. 2018. *Rise and Kill First: The Secret History of Israel's Targeted Assassinations.* New York: Random House.
- Broad, William J., John Markoff, and David E. Sanger. 2011. "Stuxnet Worm Used Against Iran Was Tested in Israel." *The New York Times,* January 15, 2011.
- Centers for Disease Control and Prevention. 2021. *History of Polio.* Atlanta, GA: U.S. Department of Health and Human Services.
- Chernow, Ron. 2010. *Washington: A Life.* New York: Penguin Press.
- Ellis, Joseph J. 2002. *Founding Brothers: The Revolutionary Generation.* New York: Vintage Books.
- Ferling, John. 2007. *Almost a Miracle: The American Victory in the War of Independence.* Oxford: Oxford University Press.
- Feynman, Richard P. 1965. *The Character of Physical Law.* Cambridge, MA: MIT Press.

- Fleming, Thomas. 1997. *Liberty! The American Revolution.* New York: Viking Penguin.
- Franklin, Benjamin. 2003. *Autobiography and Other Writings.* New York: Penguin Classics.
- Global Polio Eradication Initiative. 2013. *Polio Eradication & Endgame Strategic Plan 2013–2018.* Geneva: World Health Organization.
- Global Polio Eradication Initiative. 2023. *Annual Report 2022.* Geneva: World Health Organization.
- Global Polio Eradication Initiative. 2025. *General Factsheet (April 2025).* Geneva: World Health Organization.
- Hayden, Michael V. 2016. *Playing to the Edge: American Intelligence in the Age of Terror.* New York: Penguin Books.
- Hibbert, Christopher. 1990. *Redcoats and Rebels: The American Revolution through British Eyes.* New York: W. W. Norton.
- Kushner, David. 2013. "The Real Story of Stuxnet." *IEEE Spectrum,* February 26, 2013.
- Langner, Ralph. 2013. *To Kill a Centrifuge: A Technical Analysis of What Stuxnet's Creators Tried to Achieve.* Hamburg: The Langner Group.
- Lindsay, Jon R. 2013. "Stuxnet and the Limits of Cyber Warfare." *Security Studies* 22 (3): 365–404. https://doi.org/10.1080/09636412.2013.816122
- Mount Vernon Ladies' Association. n.d. "Committees of Correspondence." *George Washington's Mount Vernon.* Mount Vernon, VA.
- Naval History and Heritage Command. n.d. "Continental Congress and the Navy." Washington, DC: U.S. Department of the Navy.
- National Park Service. 2024. "Battle of the Capes." Yorktown, VA: Colonial National Historical Park.
- National Park Service. 2024. "Events Leading to the Siege of Yorktown." Yorktown, VA: Colonial National Historical Park.

- Our American Revolution (Jamestown-Yorktown Foundation). n.d. "February 27, 1782: Parliament Calls for End to Offensive Operations in America." Williamsburg, VA.
- Paine, Thomas. 1776. *Common Sense.* Philadelphia.
- Philbrick, Nathaniel. 2018. *In the Hurricane's Eye: The Genius of George Washington and the Victory at Yorktown.* New York: Viking.
- Rakove, Jack N. 2010. *Revolutionaries: A New History of the Invention of America.* New York: Houghton Mifflin Harcourt.
- Rose, Alexander. 2007. *Washington's Spies: The Story of America's First Spy Ring.* New York: Bantam Books.
- Rotary International. 2020. *PolioPlus: Rotary's Role in Eradication.* Evanston, IL: Rotary International.
- Salk, Jonas. 1972. *Man Unfolding.* New York: Harper & Row.
- Sanger, David E. 2012. *Confront and Conceal: Obama's Secret Wars and Surprising Use of American Power.* New York: Crown.
- Steuben, Friedrich Wilhelm von. 1779. *Regulations for the Order and Discipline of the Troops of the United States.* Philadelphia: Styner and Cist.
- Symantec (Nicolas Falliere, Liam O. Murchu, and Eric Chien). 2011. *W32.Stuxnet Dossier.* Mountain View, CA: Symantec Security Response.
- Symantec (Geoff McDonald, Liam O. Murchu, Stephen Doherty, and Eric Chien). 2013. *Stuxnet 0.5: The Missing Link.* Mountain View, CA: Symantec Security Response.
- Thompson, Kimberly M., and Radboud J. Duintjer Tebbens. 2014. "Lessons from the Polio Endgame: Overcoming the Final Hurdles." *Journal of Infectious Diseases* 210 (S1): S475–S483. https://doi.org/10.1093/infdis/jiu447
- U.S. Department of the Treasury. n.d. "Treasurer of the United States." Washington, DC.

- United States Army. n.d. "Birth of the U.S. Army (June 14, 1775)." Washington, DC: Office of the Chief of Military History.
- Wood, Gordon S. 2002. *The American Revolution: A History.* New York: Modern Library.
- World Health Organization. 1988. *World Health Assembly Resolution WHA 41.28: "Global Eradication of Poliomyelitis by the Year 2000."* Geneva: WHO.
- World Health Organization. 2021. *Polio Eradication Strategy 2022–2026: Delivering on a Promise.* Geneva: WHO.
- World Health Organization. 2023. *Field Guidance for the Implementation of Environmental Surveillance for Poliovirus.* Geneva: WHO.
- World Health Organization. 2025. "Statement of the Forty-Second Meeting of the Polio IHR Emergency Committee under the International Health Regulations (2005)." Geneva: WHO.
- World Health Organization, South-East Asia Regional Office. 2014. "WHO South-East Asia Region Officially Certified Polio-Free." New Delhi: WHO SEARO.
- Wright, Robert K. Jr. 1983. *The Continental Army.* Washington, DC: Center of Military History, United States Army.
- Zetter, Kim. 2014. *Countdown to Zero Day: Stuxnet and the Launch of the World's First Digital Weapon.* New York: Crown.

# CONCLUSION
## A CALL TO ACTION

*"Knowing is not enough; we must apply. Being willing is not enough; we must do."* — Leonardo da Vinci

You've reached the end of a book about conflict. But it wasn't a book about one kind of conflict or one kind of domain. It was a book about the structure, patterns, and strategies beneath them all.

You've seen how wars are launched and how malware spreads. How corporate sabotage mirrors battlefield tactics. How systems collapse when trust erodes—and how they hold when posture, timing, and clarity are strong. You've seen football teams and protest movements, bad actors and good intentions, victory and failure, collapse and recovery. And through it all, you've seen the same underlying principles surface again and again.

This book wasn't about mastering a single craft. It was about learning to *see*. To notice the shared dynamics that govern struggle— across spaces, across scales, and across disciplines. To speak the grammar of competition, crisis, resistance, and survival in a common tongue. UTAD is not a formula. It is a lens. And once you've seen through it, the world doesn't look the same.

Because now, you don't just see events. You see structure.

With that vision comes responsibility—strategic knowledge is never neutral.

It can be used to protect or to control, to stabilize or to dominate, to de-escalate or to strike first. The same principle that helps people withstand pressure can be used to apply it. The same techniques that help prevent collapse can also help accelerate it.

What matters is not just *what* you know—but *how* you choose to apply it. That's the responsibility this book leaves with you.

A framework like UTAD can't decide who's right. It can't tell you which side of a conflict is just. It doesn't deliver moral answers. It reveals how struggle works—before, during, and after it happens. It shows you where systems break, how they hold, and what it takes to survive with your integrity intact.

And once you see those patterns, you can't unsee them. That clarity is power. But it's also a risk. A strategist who prepares deeply might become obsessed with control. One who studies escalation might begin to see every disagreement as a threat. The same tools that make someone effective under pressure can also make them dangerous. The danger isn't knowledge—it's detachment. The most harmful actors are often those who understand the field of play but feel no responsibility for what happens inside it.

If you've come this far, you're likely someone who carries responsibility. Maybe you lead, analyze, defend, respond, teach, or build. Maybe you've already felt the weight of decisions made under stress. If so, you've been trusted with a boundary—and UTAD gives you tools for protecting it.

This framework doesn't belong to one domain. It's not about becoming a master of war, or cyber, or law enforcement. It's about understanding how conflict unfolds across all of them—and how people, systems, and adversaries behave under pressure. Whether you're managing a crisis, facing competition, defending infrastructure, or simply trying to build something that lasts, UTAD helps you act deliberately—not reactively.

You now know how to spot early warning signs. How to time your moves. How to shape perception, adapt faster, and recover stronger. Those are *real* skills. They don't guarantee victory. But they do help you avoid mistakes that lead to collapse and increase your odds of success.

And they're not just for high-stakes arenas:

- You can shape the environment before a hard conversation.
- You can gather intelligence before a job interview.
- You can make a tactical sacrifice to start a side business.

Regardless of stakes, the adversary won't always be wearing a military uniform or a martial arts *gi*. But the patterns still hold. And if you apply them with care—not perfection, but intention—they'll help you hold, too.

This wasn't a book about theory—not really—it was a map. A map of the game board of conflict. And now you understand the game better than most ever will. That puts you in a unique position to win. In your own arena. In your own life.

Your move.

# APPENDIX
## UTAD PRINCIPLES IN SUMMARY

1. **Prevent Conflict Proactively:** The most effective victory is avoiding a fight altogether by proactively addressing the root causes of conflict. This involves building strong relationships, aligning incentives, and detecting grievances early to create an environment where aggression is irrational. By investing in prevention, you can neutralize threats before they fully form, saving immense costs down the line.

2. **Define Victory with Precision:** To avoid aimless conflict and wasted resources, you must define a clear, measurable, and achievable end state before the engagement begins. This precise definition of "victory" serves as a guiding principle for all strategic and tactical decisions, ensuring that every action is purposeful. Without such clarity, missions are prone to scope creep, emotional decision-making, and an inability to recognize when the conflict is truly over.

3. **Build Capability That Renders Strategy Irrelevant:** While clever strategies are valuable, they are most decisive only when opponents are evenly matched. The ultimate

advantage comes from developing such overwhelming
capability—in talent, training, and tools—that the
adversary's strategic planning becomes futile. True
dominance is achieved not in the moment of conflict, but
through the relentless, behind-the-scenes work of
building superior strength.

4. **Know the Adversary Better Than They Know You:**
Gaining an information advantage by deeply
understanding your adversary's motivations, habits, and
vulnerabilities is crucial for success. This requires a
disciplined cycle of intelligence gathering, analysis, and
prediction, while simultaneously concealing your own
intentions and capabilities. The side that can see the
board more clearly can anticipate moves, set traps, and
shape the conflict to its advantage.

5. **Align Internally:** An organization's effectiveness in a
conflict is determined by its internal cohesion and the
alignment of its purpose, objectives, incentives, and
routines. Misalignment creates internal friction, slows
down decision-making, and opens up vulnerabilities that
an adversary can exploit. True strength in an adversarial
dynamic comes from a unified and synchronized internal
system where all parts work together seamlessly.

6. **Ally with Purpose:** Alliances should be built on a
foundation of shared, long-term purpose and
complementary capabilities, not just on the convenience
of a mutual enemy. Successful coalitions require clear
commitment mechanisms, continuous communication,
and the ability to adapt together as the environment
changes. A well-chosen ally can be a powerful force
multiplier, but a poorly aligned one can become a
strategic liability.

7. **Plan for All Scenarios:** While no plan survives contact
with the enemy, the process of planning is essential for
building the mental and logistical readiness to handle

chaos. By systematically considering a range of potential futures—primary, alternate, contingency, and emergency (PACE)—an organization can develop the flexibility to pivot effectively when faced with unexpected events. The goal is not to predict the future perfectly, but to cultivate a culture of preparedness that can withstand shocks.

8. **Test Against Realistic Resistance:** True capability can only be validated by testing it against an adversary that is actively and adaptively trying to win. Training against compliant or predictable opponents creates a false sense of security and leaves systems vulnerable to the chaos of a real-world engagement. By embracing rigorous, uncooperative testing, an organization can identify and fix weaknesses before they become fatal.

9. **Account for Insider Risk:** The most significant threats to an organization often come from within, whether through malicious intent, manipulation, mistake, or marginalization. A comprehensive defense requires a layered approach that includes people, processes, and platforms to detect and mitigate these internal risks. Ultimately, a strong, positive culture is the most effective safeguard against insider threats.

10. **Engage Stakeholders with Intentionality:** In any conflict, there are influential parties who are not direct combatants but whose support is critical for success. These stakeholders—be they regulators, investors, or community leaders—must be intentionally and proactively managed through clear communication and an understanding of their wants and fears. Failing to secure their buy-in can lead to a loss of legitimacy and resources, dooming an otherwise sound strategy.

11. **Shape the Environment to Your Advantage:** Before a conflict even begins, a skilled strategist works to tilt the playing field in their favor. This involves manipulating the physical, procedural, informational, and psychological

"terrain" to create advantages that will pay dividends later. By shaping the environment, you can constrain your adversary's options and make your own path to victory smoother.

12. **Strike First and Sustain Pressure:** When conflict is inevitable, seizing the initiative with a well-timed first move can be a decisive advantage. This initial strike should be followed by relentless, sustained pressure that prevents the adversary from recovering and regaining their footing. The goal is to keep the opponent off-balance and reacting to your moves, thereby controlling the tempo of the engagement.

13. **Target Weakness and Avoid Strength:** The most efficient path to victory is to avoid a direct confrontation with an adversary's strengths and instead focus your attacks on their vulnerabilities. This requires a deep understanding of the opponent's dependencies, blind spots, and internal frictions. By applying pressure where they are least prepared, you can achieve disproportionate results with minimal effort.

14. **Reduce Your Exposure:** To be a difficult target, you must limit what your adversary can see, attack, and damage. This involves a dynamic approach to concealment, dispersion of critical assets, and the segmentation of systems to minimize the impact of any single breach. By making yourself a more elusive and resilient target, you increase the cost and effort for your adversary to engage you effectively.

15. **Exploit the Environment:** A masterful strategist does not just operate within their environment; they use its existing features as a weapon. This involves leveraging the physical terrain, procedural rules, or informational landscape to constrain an opponent and amplify one's own strengths. By turning the environment into an active ally, even a weaker force can gain a significant advantage.

16. **Time Your Actions Strategically:** The effectiveness of an action is often determined more by when it is taken than by what is done. Strategic timing requires relentless preparation, the ability to sense the critical moment of opportunity (*kairos*), and the skill to orchestrate the tempo of your actions. A well-timed move can catch an adversary off guard and create a decisive opening, while a poorly timed one can lead to failure.

17. **Adapt Quickly to Changing Conditions:** In a dynamic adversarial environment, the ability to rapidly detect and respond to change is paramount for survival. Rigid systems that are slow to adapt become predictable and vulnerable. True adaptability requires a culture and structure that support early detection, accelerated decision-making, and the flexible reallocation of resources.

18. **Stretch the Adversary's Defenses:** An adversary with finite resources can be weakened by forcing them to defend multiple fronts simultaneously. By creating threats across different spaces, modes, and times, you can dilute their focus and energy, making them vulnerable. A stretched defense is a fragile defense, and it creates opportunities for a decisive breakthrough.

19. **Deceive to Force Mistakes:** All warfare is based on deception, and a well-crafted ruse can lead an adversary to misallocate resources, misjudge intentions, and ultimately defeat themselves. Effective deception requires a deep understanding of the adversary's psychology and a meticulous five-step process: design, planting, reinforcement, monitoring, and adaptation. By controlling what the enemy sees, you can control what they do.

20. **Sacrifice Tactically to Gain Strategically:** A powerful, though counterintuitive, strategy is to intentionally accept a short-term loss to achieve a larger, long-term gain. This is not a retreat, but a purposeful and asymmetrical trade of

a smaller piece for a more valuable strategic position. A well-executed sacrifice can disarm an opponent, seize the initiative, and create opportunities that would not otherwise exist.

21. **Drain the Adversary's Resources Systematically:** Victory can be achieved not just by outmaneuvering an opponent, but by outlasting them. A strategy of attrition aims to systematically drain the adversary's material, cognitive, and emotional resources at a rate faster than they can be replenished. By imposing disproportionate costs on the enemy, you can lead them to a state of exhaustion and collapse.

22. **Disrupt the Adversary's Control Center:** Every system has a nerve center—a key node responsible for decision-making and coordination. By targeting and disrupting this control center, you can create paralysis and fragmentation throughout the adversary's entire network. This approach is highly efficient, as a single, surgical strike on the "brain" can neutralize the entire "body."

23. **Break the Adversary's Will to Fight:** The most profound victories are won not on the battlefield, but in the mind of the adversary. By targeting the psychological pillars of their resistance—morale, cohesion, and belief in victory— you can dissolve their will to continue the fight. When an adversary no longer sees a purpose in fighting, physical confrontation becomes unnecessary.

24. **Calculate Risk and Reward in Every Action:** Every decision in a conflict should be the result of a disciplined and repeatable process of weighing potential rewards against potential costs and probabilities of success. This requires a clear-eyed assessment of all factors, including intangible ones like morale and political capital, and a commitment to data-driven decision-making over pure instinct. A sound risk-reward calculus is the engine of effective strategy.

25. **Maintain Composure Under Pressure:** The immense pressure of a conflict can sabotage judgment and lead to disastrous, fear-driven decisions. Maintaining composure is a trainable skill, cultivated through stress-inoculation drills and anchored by a clear, unwavering mission or credo. A leader's calm under pressure is contagious and can be the deciding factor between a disciplined response and a panicked collapse.

26. **Nurture Your Critical Resources:** The long-term success of any endeavor depends on the daily, unglamorous work of caring for your most critical resources: your people, your equipment, and your intangible capital like trust and reputation. This is about pragmatic stewardship, ensuring that your "ship" is sound and your "crew" is ready before the storm hits. A well-nurtured force is a resilient and enduring one.

27. **Pace Your Actions for Sustainability:** In any long-term conflict, endurance is more important than initial speed. A sustainable pace involves calibrating your efforts to the expected duration of the engagement, managing your physical, cognitive, emotional, and systemic energy reserves. The goal is to avoid burnout and remain effective for the entire race, not just the opening sprint.

28. **Uphold Your Integrity:** In the heat of a conflict, there is always a temptation to sacrifice your values for a tactical advantage. However, integrity is not a luxury; it is a strategic asset that fosters trust, cohesion, and long-term viability. By defining and adhering to your non-negotiables, you protect your identity and ensure that a victory is not won at the cost of your soul.

29. **Finish Smartly in Victory or Defeat:** The end of a conflict is as critical as the beginning and requires overcoming the emotional distortions of both overconfidence in victory and stubbornness in defeat. A smart finish involves locking in a win with continued focus rather than

coasting, or executing a clean, strategic withdrawal from a losing position to preserve resources for the next engagement. The goal is to close the current conflict in a way that best positions you for the future.

30. **Regroup, Learn, and Rebuild:** The true conclusion of any conflict comes only after you have systematically learned from the experience and translated those lessons into improved capabilities. This three-phase process—regrouping to stabilize, learning with ego-less honesty, and rebuilding to create lasting change—is the engine of strategic evolution. By ensuring that you never face an adversary with the same vulnerabilities twice, you turn every engagement, win or lose, into a source of strength.

# ABOUT THE AUTHOR

Chris Griggs is an intelligence professional, cybersecurity leader, and educator with nearly three decades of experience across the defense, law enforcement, education, and private sectors. His career spans counterterrorism, cyber threat intelligence, technical presales, training development, and cross-domain analysis—anchored by a rare blend of national security insight and systems-level thinking.

A former U.S. Army Counterintelligence Special Agent, Chris deployed in support of Operations Enduring Freedom and Iraqi Freedom, later serving as an Intelligence Analyst at U.S. Central Command and U.S. Special Operations Command, and revolutionized the Florida Highway Patrol's threat intelligence capabilities during his tenure as Chief Intelligence Officer. In the private sector, he currently leads technical solutions architecture for national security clients at Flashpoint, specializing in cyber/OSINT.

Chris is also an adjunct professor at Florida State University, where he has helped shape the next generation of intelligence professionals since 2014. He holds a Bachelor of Arts in International Studies from the University of South Florida and a Master of Science in Criminology from Florida State University. He is currently pursuing a Specialist in Information degree (focused on artificial intelligence, trust and safety, and cyber threat intelligence) also from Florida State University. His research interests include emerging technologies, the future of AI, and the dynamics of conflict.

Chris lives in Tallahassee, Florida, with his wife and two children.

# GLOSSARY

- **Adaptive Force:** Any agent or system capable of changing its behavior in response to pressure or resistance.
- **Adversarial Configuration:** The structural makeup of an agent or force within a conflict, which can include Single Agent, Coordinated Group, Distributed Network, and Fragmented System.
- **Adversarial Dynamics:** The study of opposition between adaptive forces or agents with incompatible goals. It also refers to the structural behavior that emerges when these forces oppose each other over time.
- **Adversarial Pairing:** A formal configuration of conflicting entities, such as Human vs. Artificial Intelligence or Animal vs. Unconscious Life.
- **Adversarial Posture:** The strategic stance an agent adopts toward a conflict, which can be offensive, defensive, hybrid, or passive-obstructive.
- **Agent:** An entity, such as a person, system, collective, or algorithm, that is capable of independent action and intention within a conflict.

- **Agent Type:** A classification of adversarial entities, categorized as Human, Animal, Artificial Intelligence, or Unconscious Life.
- **Animal vs. Animal:** An adversarial domain examining conflict between or within species, which is often driven by instinct or survival.
- **Artificial Intelligence vs. Artificial Intelligence:** A conflict domain centered on algorithmic or machine-led conflict without direct human oversight.
- **Asymmetry:** A mismatch in capabilities, constraints, configuration, goals, resources, or structure between adversaries.
- **Attack Surface:** The total of all physical, digital, psychological, or procedural vulnerabilities that an agent or system presents to an adversary.
- **Attritional Pressure:** The deliberate use of sustained material or psychological costs to exhaust an adversary.
- **Center of Gravity:** The critical source of power, lynchpin, or node that provides moral or physical strength and holds an adversary's system together.
- **Chokepoint:** A narrow corridor or resource dependency that can be used to control or constrain an adversary's movement.
- **Cognitive Load:** The total mental processing demand placed on individuals or teams during complex or high-stakes situations.
- **Command-and-Control (C2):** The central or distributed mechanism that allows a force to make unified decisions and act in a coordinated manner.
- **Composure Failure:** A breakdown in discipline, clarity, or judgment caused by fear, stress, or emotional escalation.
- **Concealment:** Hiding capabilities, intentions, or location to reduce an adversary's ability to see or target you, which is distinct from deception.

- **Constraint Environment:** The set of normative or regulatory boundaries (legal, moral, procedural, or none) within which an agent operates.
- **Control:** A victory condition defined by achieving influence or command over an adversary, domain, or outcome without necessarily destroying it.
- **Control Node:** A key system, leader, or decision point that, if disabled, disrupts an adversary's coherence.
- **Cost-Imposition Ratio:** A metric that compares the resources an actor expends to the cost inflicted on their adversary.
- **Cultural Terrain:** The informal values, expectations, or assumptions within a group that shape perception and behavior.
- **Deception:** A tactic used to manipulate an adversary's perception or behavior.
- **Decision Fatigue:** Cognitive exhaustion resulting from repeated decision-making under stress or uncertainty.
- **Displacement Risk:** The strategic danger of misinterpreting adaptive changes as threats instead of reorganizations.
- **Distributed Network:** A configuration of multiple semi-autonomous agents with shared goals but limited central control, which often results in emergent behavior and resilience.
- **Doctor vs. Pathology:** An adversarial domain examining the conflict between a healer and a condition, focused on diagnosis, intervention, and adaptation.
- **Elimination:** A victory condition aiming for the total removal, destruction, or incapacitation of the adversary.
- **Engagement Phase:** The active period of an adversarial interaction, including maneuvers, attacks, defense, and real-time adaptation.
- **Entropy (Strategic):** The gradual decline of order or

capability within a system, which an adversary can exploit or accelerate.

- **Environmental Exploitation:** The tactical use of terrain, infrastructure, or system architecture to constrain an adversary.
- **Escape:** A victory condition focused on withdrawal or evasion rather than confrontation.
- **Ethical Collapse:** A strategic failure where an entity abandons its moral principles to win, leading to a loss of integrity.
- **Ethical Red Line:** A predefined limit that an actor will not cross, regardless of the tactical advantage it might offer.
- **Force Multiplier:** A factor that gives an entity a disproportionately large increase in effectiveness.
- **Fragmented System:** A configuration that is internally incoherent or lacks unity, making it resistant to control.
- **Friction (Strategic):** The cumulative disruption of coordination, tempo, or clarity within a system caused by internal or external stressors.
- **Human vs. Artificial Intelligence:** A domain where humans oppose autonomous machine systems.
- **Human vs. Bureaucratic Resistance:** A domain where individuals face systems that resist change through complexity, inertia, or passive opposition.
- **Human vs. Human:** A domain of direct interpersonal or intergroup conflict.
- **Human vs. Institution:** A domain where individuals resist structured entities like governments or corporations.
- **Human vs. Self:** A domain of internal conflict against parts of oneself, like habits or doubts.
- **Human vs. System:** A domain where humans engage adversarially with non-sentient systems like policies or algorithms.
- **Human vs. Unconscious Life:** A domain involving conflict

with non-sentient biological systems like viruses, where resistance is real but lacks intent.

- **Information Asymmetry:** A structural imbalance in knowledge or situational awareness between opposing forces.
- **Initiative Seizure:** The act of gaining and maintaining control over the tempo and direction of a conflict.
- **Institutional Memory:** The accumulated knowledge within a system, often embedded in its doctrine or norms.
- **Intelligence Loop:** The continuous cycle of collection, analysis, decision-making, and feedback that guides adaptation.
- **Internal Coherence:** The alignment of goals, behaviors, and incentives within a system.
- **Invasive Species:** A domain studying non-native organisms entering an environment, often used as an analogy for other types of intrusions.
- **Invisible Drain:** The hidden depletion of morale, clarity, or stamina over time.
- **Kill-Switch Threshold:** A predefined trigger that halts an action or initiates a fallback due to unacceptable risk.
- **Latency Window:** The time delay between a change in conditions and an actor's effective response.
- **Legal/Rules-Based Constraint:** A constraint environment where codified laws or formal regulations govern actions.
- **Legitimacy Erosion:** The decline of internal or external trust in an actor's moral or legal right to act.
- **Logistics Pulse:** The rhythm and reliability of resource flow within a system during an active engagement.
- **Mirror-Imaging:** A cognitive error where one projects their own values or assumptions onto an adversary.
- **Mission Creep:** The tendency for a project to expand beyond its original goals, leading to strategic drift.
- **Moral Coherence:** The alignment of an entity's actions

with its stated values, which is a key component of integrity.

- **Moral Load:** The psychological burden carried by individuals or systems under sustained ethical compromise.
- **Narrative Terrain:** The informational space where legitimacy, support, and perception are contested.
- **Network Collapse:** The sudden disintegration of a distributed system.
- **Normalcy Bias:** The psychological tendency to underestimate the potential for significant disruption.
- **OODA Loop:** A four-stage decision-making model (Observe, Orient, Decide, Act) where speed and effectiveness provide a decisive advantage.
- **Operational Drag:** The resistance from bureaucratic delay, unclear incentives, or unaligned priorities.
- **Opportunity Cost (Adversarial):** The strategic value of a path not taken during an engagement.
- **Optimization:** A victory condition defined by improving performance or efficiency.
- **Postengagement Phase:** The period after active conflict, including assessment, recovery, and learning.
- **Preengagement Phase:** The period before conflict, involving scanning, positioning, and shaping actions.
- **Principle (UTAD):** A fundamental rule about adversarial behavior.
- **Psychological Fragmentation:** The breakdown of unified purpose or belief in victory within a force.
- **Red Team / Blue Team:** A simulation of adversarial conflict between offensive (red) and defensive (blue) teams, typically in cybersecurity.
- **Redundancy Design:** The deliberate creation of backup capabilities to maintain function under pressure.
- **Reflex Collapse:** The failure of automated or default behaviors under new or extreme conditions.

- **Resilience:** The capacity to endure, recover, or adapt during adversarial pressure without collapsing.
- **Resolution:** A victory condition where the goal is mutual understanding, de-escalation, or a sustainable compromise.
- **Resource Gradient:** The difference in resource consumption rates between opposing forces.
- **Resonance (Strategic):** The alignment between an action and a narrative that generates momentum or legitimacy.
- **Risk Envelope:** The boundary of acceptable uncertainty or exposure in an engagement.
- **Signal Pollution:** The presence of conflicting or excessive information that obscures actionable intelligence.
- **Single Agent:** A configuration where one actor functions independently in a conflict.
- **Strategic Drift:** The unintentional changing or loss of a primary objective due to pressure or cumulative deviation.
- **Strategic Initiative:** The ability to control the tempo, terms, and direction of an engagement.
- **Stress Inoculation:** Building psychological resilience by exposing individuals to controlled pressure in training.
- **Structural Vulnerability:** A weakness embedded in the architecture of a system that is often invisible until exploited.
- **Sunk Cost Fallacy:** Continuing a failing action because of invested resources rather than future prospects.
- **Tempo Control:** The ability to dictate the speed and rhythm of action during an engagement.
- **Temporal Phases:** The three main stages of an adversarial engagement: Preengagement, Engagement, and Postengagement.
- **Threat Surface:** The total area—physical, digital, or cognitive—on which an adversary can apply pressure.
- **Turning Radius:** The time or effort required for a system to change direction once committed.

- **A Unified Theory of Adversarial Dynamics (UTAD):** A framework that maps the shared principles of struggle across numerous domains to provide a universal grammar for understanding conflict.
- **Unconscious Life vs. Unconscious Life:** A domain where biological organisms resist or displace each other without consciousness or intent.
- **Unforced Error:** A costly mistake made without direct adversary pressure.
- **Value Anchor:** A guiding belief or purpose that orients action during chaos.
- **Victory Condition:** The explicit goal or criteria that constitutes success in an engagement. UTAD defines seven: Elimination, Control, Escape, Survival, Resolution, Decision, and Optimization.
- **Will Collapse:** The point at which an adversary no longer believes continuing is worthwhile, regardless of their remaining capacity.

# BIBLIOGRAPHY

## Works Cited and Referenced

- A&M Records, Inc. v. Napster, Inc., 239 F.3d 1004 (9th Cir. 2001).
- Abraham, Thomas. 2018. *Polio: The Odyssey of Eradication*. New York: Columbia University Press.
- Adams Papers Digital Edition (Massachusetts Historical Society). 1782/2018. *Correspondence Referencing the House of Commons Debate and Motion of 27 February 1782 to Discontinue Offensive War in America*. Boston: MHS.
- Aesop. 1998. *Aesop: The Complete Fables*. Translated by Olivia and Robert Temple. London: Penguin Classics.
- Aesop. n.d. "The Tortoise and the Hare." In *Aesop's Fables*.
- Akins, Joyce. 2010. "The Falklands War: A Study in Crisis Management and Public Diplomacy." *Naval War College Review* 63, no. 3: 97–116.
- Ambrose, Stephen E. 1994. *D-Day: June 6, 1944*. New York: Simon & Schuster.
- Amazon Web Services. 2017. "GameDay: Simulating Real-World Outages to Build Resiliency." AWS Architecture Blog, November 14. http://aws.amazon.com/blogs/architecture/gameday-simulating-real-world-outages-to-build-resiliency/
- Apple. 2005. "Apple to Use Intel Microprocessors Beginning in 2006." News release, June 6. http://www.apple.com/newsroom/2005/06/06Apple-to-Use-Intel-Microprocessors-Beginning-in-2006/
- Argyris, Chris. 1990. *Overcoming Organizational Defenses: Facilitating Organizational Learning*. Boston: Allyn & Bacon.
- Associated Press. 2017. "Case in Fatal Shooting of Kmart Shoplifter Settled for $285K." *The Washington Times*, April 20.
- The Atlantic. 2020. "How the Pandemic Defeated America." *The Atlantic*, April.
- Axelrod, Robert. 1984. *The Evolution of Cooperation*. New York: Basic Books.
- Aylward, Bruce. 2012. Interview by Rotary International. Video, May.
- Bailyn, Bernard. 1967. *The Ideological Origins of the American Revolution*. Cambridge, MA: Harvard University Press.

- Ball, Deborah, and Daniel Levitt. 2023. "What Was the Cost of the Lockdowns?" *The Wall Street Journal*, March 10.
- Bank of England. 1993. *Report on Withdrawal from the Exchange Rate Mechanism*. London: HMSO.
- Bar-Joseph, Uri. 2005. *The Watchman Fell Asleep: The Surprise of Yom Kippur and Its Sources*. Albany: State University of New York Press.
- Baseball Prospectus (Cot's Baseball Contracts). n.d. *Opening Day Payrolls, 2002*. Baseball Prospectus.
- Baseball-Reference. n.d. *Chad Bradford Statistics*. Sports Reference LLC.
- Baseball-Reference. n.d. *David Justice Statistics*. Sports Reference LLC.
- Baseball-Reference. n.d. *Oakland Athletics 2002 Game Log*. Sports Reference LLC.
- Baseball-Reference. n.d. *Scott Hatteberg Statistics*. Sports Reference LLC.
- Baseball-Reference. n.d. *2002 Oakland Athletics Team Page/Statistics*. Sports Reference LLC.
- BBC News. 2007. "Timeline: The Falklands Conflict." BBC News, April 2.
- Beilock, Sian. 2010. *Choke: What the Secrets of the Brain Reveal About Getting It Right When You Have To*. New York: Free Press.
- Bendavid, Eran, Christopher Oh, Jay Bhattacharya, and John P. A. Ioannidis. 2021. "Assessing Mandatory Stay-at-Home and Business Closure Effects on the Spread of COVID-19." *European Journal of Clinical Investigation* 51 (4): e13484. https://doi.org/10.1111/eci.13484
- Bergman, Ronen. 2018. *Rise and Kill First: The Secret History of Israel's Targeted Assassinations*. New York: Random House.
- bin Ladin, Usama. 2004. "Full Transcript of bin Laden's Speech." *The Guardian*, October 29.
- BizTech Magazine. 2017. *How Data Analytics Is Changing Baseball*. Arlington, VA: BizTech Magazine.
- Bloomberg Businessweek. 2012. "Tylenol Survivor: How J&J Triumphed After Tragedy." *Bloomberg Businessweek*, September.
- Bonaparte, Napoléon. 1966. Quoted in J. L. Heinl Jr., ed. *Dictionary of Military and Naval Quotations*. Annapolis, MD: United States Naval Institute.
- Bowden, Mark. 2001. *Killing Pablo: The Hunt for the World's Greatest Outlaw*. New York: Atlantic Monthly Press.
- BoxRec. n.d. "Mike Tyson." http://boxrec.com/en/proboxer/474
- Boyd, John. 1986. "A Discourse on Winning and Losing." Lecture notes, Maxwell AFB, AL.
- Boyd, John. 1986. *Patterns of Conflict*. Briefing slides.

- Brauner, Jan M., et al. 2021. "Inferring the Effectiveness of Government Interventions against COVID-19." *Science* 371 (6531): eabd9338. https://doi.org/10.1126/science.abd9338

- Britannica. n.d. "Falkland Islands War." *Encyclopedia Britannica*.

- Broad, William J., John Markoff, and David E. Sanger. 2011. "Stuxnet Worm Used Against Iran Was Tested in Israel." *The New York Times*, January 15.

- Bromiley, Matt. 2023. "Know Thyself, Know Thy Enemy: A Proactive Approach to External Attack Surface Management." SANS Institute Whitepaper.

- Brown, Judith M. 1977. *Gandhi and Civil Disobedience: The Mahatma in Indian Politics, 1928–1934*. Cambridge: Cambridge University Press.

- *Bringing Down a Dictator*. 2002. Directed by Steve York. Washington, DC: York Zimmerman Inc. (PBS broadcast).

- Bshary, Redouan, Andrea Hohner, Karim Ait-el-Djoudi, and Hans Fricke. 2006. "Interspecific Communicative and Coordinated Hunting between Groupers and Giant Moray Eels in the Red Sea." *PLOS Biology* 4, no. 12: e431. https://doi.org/10.1371/journal.pbio.0040431

- Buckminster Fuller, R. 1981. *Critical Path*. New York: St. Martin's Press.

- Buffett, Warren. 2007. *Annual Letter to Berkshire Hathaway Shareholders*. Omaha, NE: Berkshire Hathaway Inc.

- Burgess, Graham, John Nunn, and John Emms. 2010. *The Mammoth Book of the World's Greatest Chess Games*. London: Robinson.

- Burson-Marsteller. 1984. *Tylenol Crisis Case Study*. Public Relations Society of America.

- Carnegie, Dale. 1936. *How to Win Friends and Influence People*. New York: Simon & Schuster.

- Carnegie Mellon University, CERT Division. 2023. *Insider Threat Center Glossary*. Pittsburgh, PA: Carnegie Mellon University.

- Carreyrou, John. 2015. "Hot Startup Theranos Has Struggled with Its Blood-Test Technology." *The Wall Street Journal*, October 16. http://www.wsj.com/articles/theranos-has-struggled-with-blood-tests-1444881901

- CBS 21 News. 2019. "Familiar Face Now in Charge at Springettsbury Township Police." July 16.

- Centers for Disease Control and Prevention. 2020. "Recommendation Regarding the Use of Cloth Face Coverings to Help Slow the Spread of COVID-19." Atlanta, GA: CDC, April 3.

- Centers for Disease Control and Prevention. 2021. "Mental Health Surveillance—United States, 2020–2021." *MMWR Supplement*. Atlanta, GA: CDC.

- Centers for Disease Control and Prevention. 2021. *History of Polio*. Atlanta, GA: U.S. Department of Health and Human Services.
- Centers for Disease Control and Prevention. 2024. *Fast Facts: Health and Economic Costs of Chronic Conditions*. Atlanta, GA: CDC.
- Central Intelligence Agency. 1995. *Ames Damage Assessment Report*. Declassified summary.
- Central Intelligence Agency. 1999. "Soviet Aircraft Losses in Afghanistan, 1984–1988." Declassified memo.
- Chancellor, Edward. 1999. *Devil Take the Hindmost: A History of Financial Speculation*. New York: Farrar, Straus and Giroux.
- Chernow, Ron. 1998. *Titan: The Life of John D. Rockefeller, Sr.* New York: Random House.
- Chernow, Ron. 2010. *Washington: A Life*. New York: Penguin Press.
- Chomsky, Noam. 1975. *Reflections on Language*. New York: Pantheon Books.
- Cicero, Marcus Tullius. 1976. *In Catilinam I–IV*. Translated by C. MacDonald. Cambridge, MA: Harvard University Press.
- Clark, Kenneth B., and Mamie P. Clark. 1947. "Racial Identification and Preference in Negro Children." *Journal of Negro Education* 19, no. 3: 341–50.
- Clausewitz, Carl von. 1976. *On War*. Edited and translated by Michael Howard and Peter Paret. Princeton, NJ: Princeton University Press.
- Clausewitz, Carl von. 1984. *On War*. Translated by Michael Howard and Peter Paret. Princeton, NJ: Princeton University Press.
- Clayton, Tim, and Phil Craig. 2000. *Finest Hour: The Battle of Britain*. London: Hodder & Stoughton.
- CNN. n.d. "Falklands War: Fast Facts." CNN.
- Congressional Budget Office. 2022. *Budgetary Effects of the 2020 CARES Act*. Washington, DC: CBO.
- Conn, David. 2021. "'A Grotesque Betrayal': How The Guardian Reported the Super League Story." *The Guardian*, April 23. http://www.theguardian.com/football/2021/apr/23/a-grotesque-betrayal-how-the-guardian-reported-the-super-league-story
- Connelly, Bill. 2019. "Blueprint for Modern Offenses." ESPN Analytics Notebook, August 20.
- Corrigan, Robert M. 2015. *Rodent Control: A Practical Guide for Pest Management Professionals*. Cleveland, OH: GIE Media.
- Council on Foreign Relations. n.d. "Timeline: The Iraq War." Council on Foreign Relations.
- Crawford, Neta C., and Catherine Lutz. 2021. *The U.S. Budgetary Costs of*

*the Post-9/11 Wars, 2001–2022.* Providence, RI: Brown University, Watson Institute, Costs of War Project.

- Cusumano, Michael A., and David B. Yoffie. 1998. *Competing on Internet Time: Lessons from Netscape and Its Battle with Microsoft.* New York: Free Press.
- Cybersecurity and Infrastructure Security Agency. 2020. "Alert AA20-352A: Mitigations for the SolarWinds Orion Supply-Chain Compromise." December.
- DeepMind. 2016. *AlphaGo: Mastering the Game of Go with Deep Neural Networks and Tree Search.* White paper. London: DeepMind Technologies.
- Department of Justice (DOJ). 2022. "Elizabeth Holmes Sentenced to More Than Eleven Years for Defrauding Investors." Press release, U.S. Attorney's Office, Northern District of California, November 18. http://www.justice.gov/usao-ndca/pr/elizabeth-holmes-sentenced-more-11-years-defrauding-theranos-investors-hundreds
- DeVore, Jennifer L., Rebecca J. Cramp, Richard J. Capon, Michael R. Crossland, and Richard Shine. 2021. "Chemical Cues from Cane Toad Eggs Attract Conspecific Tadpoles: A Basis for Species-Specific Trapping." *Proceedings of the National Academy of Sciences* 118 (24): e2024296118. https://doi.org/10.1073/pnas.2024296118
- Doran, George T. 1981. "There's a S.M.A.R.T. Way to Write Management's Goals and Objectives." *Management Review* 70, no. 11: 35–36.
- Douglas, Buster, and Randy Gordon. 2018. *42 to 1: The Story of the Biggest Upset in Boxing History.* B&G Publishing.
- Drug Policy Alliance. 2015. "The War on Drugs, A Trillion Dollar Failure." News release, January 8. http://drugpolicy.org/news/2015/01/war-drugs-trillion-dollar-failure
- Eastman Kodak Company. 1975–2011. *Annual Reports.* Rochester, NY: Eastman Kodak Company.
- Eastridge, Brian J., Donald Jenkins, Stephen Flaherty, Henry Schiller, and John B. Holcomb. 2006. "Trauma System Development in a Theater of War: Experiences from Operation Iraqi Freedom and Operation Enduring Freedom." *Journal of Trauma: Injury, Infection, and Critical Care* 61 (6): 1366–72. https://doi.org/10.1097/01.ta.0000245894.78941.90
- *The Economist.* 2006. "Enron's Legacy." *The Economist,* January.
- Edmondson, Amy C. 2018. *The Fearless Organization: Creating Psychological Safety in the Workplace for Learning, Innovation, and Growth.* Hoboken, NJ: Wiley.
- Eichenwald, Kurt. 2005. *Conspiracy of Fools: A True Story.* New York: Broadway Books.

- Ellis, Joseph J. 2002. *Founding Brothers: The Revolutionary Generation*. New York: Vintage Books.
- Elop, Stephen. 2011. "Burning Platform." Internal memo, Nokia Corporation, February.
- "Enron Scandal." 2002. *PBS Frontline: The Corporate Scandal Sheet*.
- ESPN. 2015. *The Great Analytics Rankings*. Bristol, CT: ESPN.
- ESPN Stats & Information. 2014. "Spurs Historically Dominant in 2014 Finals." ESPN, June 16. http://www.espn.com/blog/statsinfo/post/_/id/91567/spurs-historically-dominant-in-2014-finals
- European Centre for Disease Prevention and Control. 2012. *Healthcare-Associated Infections: Annual Epidemiological Report 2012*. Stockholm: ECDC.
- "Falkland Islands – The British Nationality Act of 1981." n.d. *Hansard*, UK Parliament.
- "Falklands War: By the Numbers." n.d. *The Telegraph*.
- Ferling, John. 2007. *Almost a Miracle: The American Victory in the War of Independence*. Oxford: Oxford University Press.
- Feynman, Richard P. 1965. *The Character of Physical Law*. Cambridge, MA: MIT Press.
- Fforde, Gervase. 2018. *Black Wednesday and the ERM Crisis*. Cambridge Economic History Papers.
- Fleming, Thomas. 1997. *Liberty! The American Revolution*. New York: Viking Penguin.
- Florance, David, John K. Webb, Thomas Dempster, Michael R. Kearney, Amy Worthing, and Mike Letnic. 2011. "Excluding Access to Freshwater by an Invasive Amphibian Can Halt Its Spread." *Proceedings of the Royal Society B* 278 (1723): 3663–70. https://doi.org/10.1098/rspb.2011.0286
- Foley, Mary Jo. 2018. "Satya Nadella's Biggest Decision: Giving Up on Mobile." *ZDNet*, July 2. http://www.zdnet.com/article/satya-nadellas-biggest-decision-giving-up-on-mobile/
- Food and Drug Administration. 1989. "Tamper-Resistant Packaging Regulations for OTC Human Drug Products; Final Rule." February 2.
- Forseth, I. N., and A. F. Innis. 2004. "Kudzu (Pueraria montana): History, Physiology, and Ecology Combine to Make a Major Ecosystem Threat." *Critical Reviews in Plant Sciences* 23, no. 5: 401–13. https://doi.org/10.1080/07352680490505150
- Franklin, Benjamin. 1735. "An Ounce of Prevention Is Worth a Pound of Cure." *Pennsylvania Gazette*, February 4.
- Franklin, Benjamin. 2003. *Autobiography and Other Writings*. New York: Penguin Classics.

- Franks, Lord. 1983. *Falkland Islands Review: Report of a Committee of Privy Counsellors*. London: HMSO, January.
- Frederick II (the Great). 1760. *Military Instructions for the Generals of His Army*.
- Freedman, Lawrence. 2005. *The Official History of the Falklands Campaign, Vol. 1: The Origins of the Falklands War*. London: Routledge.
- Freedman, Lawrence. 2013. *Strategy: A History*. New York: Oxford University Press.
- Freeman, Mike. 1999. "In Bizarre Ending, Tyson Bout Is Declared a No Contest." *The New York Times*, October 24. http://www.nytimes.com/1999/10/24/sports/in-bizarre-ending-tyson-bout-is-declared-a-no-contest.html
- Frieser, Karl-Heinz. 2005. *The Blitzkrieg Legend: The 1940 Campaign in the West*. Annapolis, MD: Naval Institute Press.
- Fritz, Martina, et al. 2011. "Linezolid-Resistant MRSA Outbreak in Intensive-Care Units, Madrid, 2008–09." *Clinical Infectious Diseases* 52, no. 5: 676–84. https://doi.org/10.1093/cid/ciq202
- Futterman, Matthew. 2011. "Masters Meltdown Still Haunts Norman." *The Wall Street Journal*, April 8.
- Gardner, Howard. 1983. *Frames of Mind: The Theory of Multiple Intelligences*. New York: Basic Books.
- Gartner. 2011. "Gartner Says Worldwide Smartphone Sales to End Users Increased 72 Percent in 2010." Press release, February 9.
- Gates, Bill. 1995. "The Internet Tidal Wave." Internal memorandum, Microsoft Corporation, May 26.
- Gawande, Atul. 2010. *The Checklist Manifesto: How to Get Things Right*. New York: Metropolitan Books.
- Gellerman, Bruce. 2011. "The Rats Are Winning." *Living on Earth*, NPR, March 11. http://www.loe.org/shows/segments.html?programID=11-P13-00010&segmentID=5
- Gentry, Justin. 2007. *The Ultimate Fighting Championship: A History of the UFC's First 100 Events*. Toronto: ECW Press.
- Global Polio Eradication Initiative. 2013. *Polio Eradication & Endgame Strategic Plan 2013–2018*. Geneva: World Health Organization.
- Global Polio Eradication Initiative. 2023. *Annual Report 2022*. Geneva: World Health Organization.
- Global Polio Eradication Initiative. 2025. *General Factsheet (April 2025)*. Geneva: World Health Organization.
- Goenner, Alec. n.d. "The Six Pillars of Krav Maga." *Israeli Krav Maga*. http://ikmakravmaga.com/resources/the-six-pillars-of-krav-maga.html

- Goldsberry, Kirk. 2019. *Sprawlball: A Visual Tour of the New Era of the NBA.* Boston: Houghton Mifflin Harcourt.
- Google. 2014. *BeyondCorp: A New Approach to Enterprise Security.* White paper. Mountain View, CA: Google.
- *Gospel of Mark.* 1982. *The New King James Version.* Nashville: Thomas Nelson.
- Gottfredson, Linda S. 1997. "Why g Matters: The Complexity of Everyday Life." *Intelligence* 24 (1): 79–132. https://doi.org/10.1016/S0160-2896(97)90014-3
- Government Accountability Office. 2024. *NASA: Assessments of Major Projects.* Washington, DC: GAO.
- Gracie, Rorion. 1994. "The Birth of Brazilian Jiu-Jitsu." Interview. *Black Belt*, July.
- Grant, Justin. 2012. "Knightmare on Wall Street: A Lesson in Risk Management." *Financial Times*, August 3.
- Grant, Ulysses S. 1885. *Personal Memoirs of U.S. Grant.* New York: Charles L. Webster & Co.
- Grau, Lester W., and Michael A. Gress. 2002. *The Soviet–Afghan War: How a Superpower Fought and Lost.* Lawrence, KS: University Press of Kansas.
- Greenberg, Andy. 2020. "The Untold Story of the Capital One Hack." *Wired*, December 15.
- Greenberg, Joel. 2014. *A Feathered River Across the Sky: The Passenger Pigeon's Flight to Extinction.* New York: Bloomsbury USA.
- Gross, Bill. 2015. "The Single Biggest Reason Why Start-Ups Succeed." TED Talk, April.
- Grossman, Dave. 1996. "Defeating the Enemy's Will: The Psychological Foundations of Maneuver Warfare." *Marine Corps Gazette.*
- Grossman, Dave, and Loren W. Christensen. 2004. *On Combat: The Psychology and Physiology of Deadly Conflict in War and in Peace.* Belleville, IL: PPCT Research Publications.
- Grynbaum, Michael M. 2007. "City Unveils New Offensive in War on Rats." *The New York Times*, July 26. http://www.nytimes.com/2007/07/26/nyregion/26rats.html
- Guderian, Heinz. 1996. *Panzer Leader.* New York: Da Capo Press.
- *The Guardian.* n.d. "The Falklands War: 30 Years On." *The Guardian.*
- Hansard. 1982. *House of Commons Debate, Falkland Islands*, April 7, vol. 21, cc961–1022. London: UK Parliament.
- Harding, Luke. 2014. *The Snowden Files: The Inside Story of the World's Most Wanted Man.* New York: Vintage Books.

- Haug, Nils, et al. 2020. "Ranking the Effectiveness of Worldwide COVID-19 Government Interventions." *Nature Human Behaviour* 4: 1303–1312. https://doi.org/10.1038/s41562-020-01009-0
- Hayden, Michael V. 2016. *Playing to the Edge: American Intelligence in the Age of Terror*. New York: Penguin Books.
- Hemingway, Ernest. 1929. *A Farewell to Arms*. New York: Charles Scribner's Sons.
- Herbert, Frank. 1965. *Dune*. Philadelphia: Chilton Books.
- Herodotus. 1996. *The Histories*. Translated by G. Rawlinson. London: Penguin Classics.
- Heuer, Richards J. 1999. *Psychology of Intelligence Analysis*. Washington, DC: Central Intelligence Agency.
- Hibbert, Christopher. 1990. *Redcoats and Rebels: The American Revolution through British Eyes*. New York: W. W. Norton.
- Holland, Tom. 2007. *Persian Fire: The First World Empire and the Battle for the West*. New York: Anchor Books.
- Hölldobler, Bert, and Edward O. Wilson. 1990. *The Ants*. Cambridge, MA: Harvard University Press.
- Holm, Holly, and Ronda Rousey. 2015. Post-fight press conference transcript, UFC 193.
- Homer. 1990. *The Iliad*. Translated by Robert Fagles. New York: Penguin Classics.
- Howard, Jeremy, Austin Huang, Zeynep Tufekci, and Trisha Greenhalgh. 2021. "An Evidence Review of Face Masks Against COVID-19." *Proceedings of the National Academy of Sciences* 118 (4): e2014564118. https://doi.org/10.1073/pnas.2014564118
- Huang, Jeff. 2014. "How Netflix Beat Blockbuster: A Business Model Innovation Study." *Harvard Business Review*, January 14.
- Hubbard, Douglas W. 2014. *How to Measure Anything: Finding the Value of Intangibles in Business*. 3rd ed. Hoboken, NJ: Wiley.
- Hume, David. 1748. *An Enquiry Concerning Human Understanding*. London: A. Millar.
- Iacoboni, Marco, et al. 2001. "Neural Mechanisms of Human Imitation and Interpersonal Interaction." *Journal of Neuroscience* 21 (19): 7767–73. https://doi.org/10.1523/JNEUROSCI.21-19-07767.2001
- IBM Security. 2023. *Cost of a Data Breach Report 2023*. Armonk, NY: IBM.
- IDC. 2014. "Smartphone OS Market Share, 2013." *IDC Quarterly Mobile Phone Tracker*, February.
- *Independence Day*. 1996. Directed by Roland Emmerich. Beverly Hills, CA: 20th Century Fox.

- International Association of Chiefs of Police. 2017. "How Small Agencies Use Data-Driven Decision-Making." Session #1962, IACP Annual Conference.
- International Military Tribunal. 1947. *Trial of the Major War Criminals before the International Military Tribunal, Nuremberg, 14 November 1945–1 October 1946.* Vol. 1 (Judgment) and Vol. 2 (Jackson's Opening). Nuremberg: IMT.
- Isaacson, Walter. 2011. *Steve Jobs.* New York: Simon & Schuster.
- Jackson, Greg. 2016. Interview on *The MMA Hour.*
- Jackson, Thomas P. 1999. *Findings of Fact.* U.S. District Court for the District of Columbia, November 5.
- Jacob Litigation. 2014. "Estate of Todd W. Shultz et al. v. Gregory T. Hadfield et al." December 17.
- Jobs, Steve. 2007. "iPhone Launch Keynote." Speech, Macworld Conference & Expo, San Francisco, January 9.
- Jones, Arthur W. n.d. Quoted phrase popular in organizational management studies.
- Kahneman, Daniel. 2011. *Thinking, Fast and Slow.* New York: Farrar, Straus and Giroux.
- Kapadia, Asif, dir. 2015. *Amy.* London: On the Corner Films / Film4.
- Kasparov, Garry. 2017. *Deep Thinking: Where Machine Intelligence Ends and Human Creativity Begins.* New York: PublicAffairs.
- Keating, Gina. 2012. *Netflixed: The Epic Battle for America's Eyeballs.* New York: Portfolio/Penguin.
- Keegan, John. 1989. *The Second World War.* New York: Penguin.
- Keegan, John. 1998. *The First World War.* New York: Vintage Books.
- Keegan, John. 2004. *Intelligence in War: Knowledge of the Enemy from Napoleon to Al-Qaeda.* New York: Vintage Books.
- Kennedy, John F. 1961. "Special Message to the Congress on Urgent National Needs." Speech, May 25.
- Kindervag, John. 2010. *Build Security into Your Network's DNA: The Zero Trust Network Architecture.* Forrester Research.
- Kluger, Richard. 2004. *Simple Justice: The History of Brown v. Board of Education and Black America's Struggle for Equality.* New York: Vintage.
- Knopper, Steve. 2009. *Appetite for Self-Destruction: The Spectacular Crash of the Record Industry in the Digital Age.* Brooklyn, NY: Soft Skull Press.
- Kolodny, Lora. 2018. "Tesla Sues Ex-Gigafactory Technician, Alleging Hacking and Sabotage." CNBC, June 21.
- Kushner, David. 2013. "The Real Story of Stuxnet." *IEEE Spectrum,* February 26.

- Kuta, Sarah. 2025. "'Robo-Bunnies' Are the Newest Weapon in the Fight Against Invasive Burmese Pythons in Florida." *Smithsonian Magazine*, July 21. http://www.smithsonianmag.com/smart-news/robo-bunnies-are-the-newest-weapon-in-the-fight-against-invasive-burmese-pythons-in-florida-180987018/
- Langner, Ralph. 2013. *To Kill a Centrifuge: A Technical Analysis of What Stuxnet's Creators Tried to Achieve*. Hamburg: The Langner Group.
- Lasker, Emanuel. 1947. *Lasker's Manual of Chess*. New York: Dover Publications.
- Lawrence, T. E. 1926. *Seven Pillars of Wisdom*. London: Jonathan Cape.
- *Le Monde*. 2015. "Microsoft accuse la plus importante perte de son histoire." *Le Monde*, July 22. http://www.lemonde.fr/economie/article/2015/07/22/microsoft-accuse-la-plus-importante-perte-de-son-histoire_4693444_3234.html
- Lever, Christopher. 2001. *The Cane Toad: The History and Ecology of a Successful Coloniser*. London: Westbury Academic.
- Lewis, Michael. 2003. *Moneyball: The Art of Winning an Unfair Game*. New York: W. W. Norton.
- Liddell, Chuck, and Chad Dundas. 2008. *Iceman: My Fighting Life*. New York: Dutton.
- Liddell Hart, B. H. 1967. *Strategy*. 2nd ed. New York: Praeger.
- Lincoln, Abraham. 1953. *The Collected Works of Abraham Lincoln*, vol. 2. Edited by Roy P. Basler. New Brunswick, NJ: Rutgers University Press.
- Lindsay, Jon R. 2013. "Stuxnet and the Limits of Cyber Warfare." *Security Studies* 22 (3): 365–404. https://doi.org/10.1080/09636412.2013.816122
- Livy. 1919. *Ab Urbe Condita*, Bk. I. Translated by B. O. Foster. Cambridge, MA: Harvard University Press.
- Lucas, Gavin. 2021. *Photographica: The Fascinating History of the Camera*. London: Bloomsbury.
- Macintyre, Ben. 2010. *Operation Mincemeat: How a Dead Man and a Bizarre Plan Fooled the Nazis and Assured an Allied Victory*. New York: Crown.
- Major League Baseball. n.d. *Qualifying Offer*. New York: MLB Advanced Media.
- Manjoo, Farhad. 2010. "Amazon's Diaper War." *Slate*, November 4.
- Marshall, Thurgood. 1952. "Argument Transcript: Brown v. Board of Education." U.S. Supreme Court, December 9.
- McGonigal, Kelly, et al. 2016. "Breath-Paced Practices Reduce Heart-Rate Reactivity Under Acute Stress." Stanford Center for Compassion and Altruism Research and Education.

- McKinsey & Company. 2012. "Delivering Large-Scale IT Projects on Time, on Budget, and on Value." August.
- McKinsey & Company. 2021. *COVID-19 and Student Learning in the United States: The Hurt Could Last a Lifetime.*
- McLean, Bethany, and Peter Elkind. 2003. *The Smartest Guys in the Room: The Amazing Rise and Scandalous Fall of Enron.* New York: Portfolio.
- Meyer, Urban, and Wayne Coffey. 2015. *Above the Line: Lessons in Leadership and Life from a Championship Program.* New York: Penguin.
- Microsoft Support. 2019. "Windows 10 Mobile End of Support: FAQ." Last updated December 10. http://support.microsoft.com/en-us/windows/windows-10-mobile-end-of-support-faq-8c6574b4-7229-5dba-7554-18c24a68250f
- Mitani, John C., David P. Watts, and Sylvia J. Amsler. 2010. "Lethal Intergroup Aggression Leads to Territorial Expansion in Wild Chimpanzees." *Current Biology* 20: R507–R508. https://doi.org/10.1016/j.cub.2010.04.021
- MIT Office of the President. 2008. "The Urgency of Doing: MIT and the Spirit of Leonardo da Vinci." MIT, August 24. http://hockfield.mit.edu/urgency-doing-mit-and-spirit-leonardo-da-vinci
- Moltke, Helmuth von. 1892. *Militärische Werke*, vol. 1. Berlin: Ernst Siegfried Mittler & Sohn.
- Mount Vernon Ladies' Association. n.d. "Committees of Correspondence." *George Washington's Mount Vernon.* Mount Vernon, VA.
- Munk, Nina. 2004. *Fools Rush In: Steve Case, Jerry Levin, and the Unmaking of AOL Time Warner.* New York: Harper Business.
- Musk, Elon. 2018. Internal email to employees, "Some Concerning News," June 18.
- Musto, David F. 1999. *The American Disease: Origins of Narcotic Control.* 3rd ed. New York: Oxford University Press.
- Nadella, Satya. 2017. *Hit Refresh: The Quest to Rediscover Microsoft's Soul and Imagine a Better Future for Everyone.* New York: Harper Business.
- NASA. 1986. *Report of the Presidential Commission on the Space Shuttle Challenger Accident.* Washington, DC: U.S. Government Printing Office.
- NASA. 2023. *Artemis Program Plan.* Washington, DC: NASA. http://www.nasa.gov/wp-content/uploads/2023/09/artemis-program-plan-2023.pdf
- NASA. n.d. "Apollo 11 Mission Overview." National Aeronautics and Space Administration. http://www.nasa.gov/mission_pages/apollo/missions/apollo11.html
- National Academies of Sciences, Engineering, and Medicine. 2017. *Pain Management and the Opioid Epidemic: Balancing Societal and Individual*

*Benefits and Risks of Prescription Opioid Use*. Washington, DC: The National Academies Press. https://doi.org/10.17226/24781

- National Basketball Association. 2012. "NBA Fines San Antonio Spurs $250,000 for Resting Players." NBA Communications, November 30. http://pr.nba.com/spurs-fined-250000-nba/
- National Commission on Terrorist Attacks Upon the United States. 2004. *The 9/11 Commission Report*. Washington, DC: U.S. Government Printing Office.
- National Football League. 2015. *NFL Game Book: Super Bowl XLIX*. February 1.
- National Institute of Standards and Technology. 2020. *SP 800-207: Zero Trust Architecture*. Gaithersburg, MD: NIST.
- National Park Service. 2024. "Battle of the Capes." Yorktown, VA: Colonial National Historical Park.
- National Park Service. 2024. "Events Leading to the Siege of Yorktown." Yorktown, VA: Colonial National Historical Park.
- NATO. n.d. *Founding Treaty Texts*. North Atlantic Treaty Organization.
- Naval History and Heritage Command. n.d. "Continental Congress and the Navy." Washington, DC: U.S. Department of the Navy.
- Netflix. n.d. "The Chaos Monkey Guide to Testing." Netflix. http://netflix.github.io/chaosmonkey/
- Newkey-Burden, Chas. 2008. *Amy Winehouse: The Biography*. London: John Blake.
- Nietzsche, Friedrich. 1966. *Beyond Good and Evil: Prelude to a Philosophy of the Future*. Translated by Walter Kaufmann. New York: Vintage Books. Originally published 1886.
- Nixon, Richard. 1971. "Remarks About an Intensified Program for Drug Abuse Prevention and Control." June 17. The American Presidency Project. http://www.presidency.ucsb.edu/documents/remarks-about-intensified-program-for-drug-abuse-prevention-and-control
- Nixon, Richard M. 1957. "Address at the National Defense Executive Reserve Conference." Speech, Washington, DC, November 14.
- Nokia Corporation. 2006. *Annual Report*. Espoo, Finland: Nokia Corporation.
- Nolan, Christopher, dir. 2017. *Dunkirk*. Burbank, CA: Warner Bros. Pictures.
- Office of Justice Programs Diagnostic Center. 2019. *Engagement Summary: Springettsbury Township Police Department 2016–2018*. Washington, DC: U.S. Department of Justice.

- Office of the Comptroller of the Currency. 2016. "Consent Order #2016-113, Wells Fargo Bank, N.A." September 8.
- Ohno, Taiichi. 1988. *Toyota Production System: Beyond Large-Scale Production*. New York: Productivity Press.
- Okamoto, Brett. 2015. "Holly Holm's Coaches Thought Ronda Rousey Would Be Aggressor at UFC 193." ESPN, November 15. http://www.espn.com/mma/story/_/id/14134847/holly-holm-coaches-thought-ronda-rousey-aggressor-ufc-193
- OpenAI. 2019. "Emergent Tool Use from Multi-Agent Interaction." OpenAI Blog, September 17.
- Osinga, Frans. 2007. *Science, Strategy and War: The Strategic Theory of John Boyd*. London: Routledge.
- Our American Revolution (Jamestown-Yorktown Foundation). n.d. "February 27, 1782: Parliament Calls for End to Offensive Operations in America." Williamsburg, VA.
- Paine, Thomas. 1776. *Common Sense*. Philadelphia.
- Panja, Tariq, and Rory Smith. 2021. "How the Super League Fell Apart." *The New York Times*, April 22. http://www.nytimes.com/2021/04/22/sports/soccer/super-league-soccer-florentino-perez.html
- Patton, George S. 1947. *War as I Knew It*. Boston: Houghton Mifflin.
- Pellegrini, John. 1999. *The Official Combat Hapkido Manual*. Boca Raton, FL: International Combat Hapkido Federation.
- Pellegrini, John. 2000. *Combat Hapkido: The Martial Art for the Modern Warrior*. Boca Raton, FL: National Self-Defense Institute.
- PennPRIME. 2018. "Department Excellence Honored at PennPRIME Conference." http://www.pennprime.com/news/department-excellence-honored-at-pennprime-conference/
- Pew Research Center. 2022. *Americans' Trust in Scientists and Other Groups Declines*. Washington, DC: Pew Research Center.
- Philbrick, Nathaniel. 2018. *In the Hurricane's Eye: The Genius of George Washington and the Victory at Yorktown*. New York: Viking.
- Phillips, Benjamin L., Gregory P. Brown, and Richard Shine. 2006. "Invasion and the Evolution of Speed in Toads." *Nature* 439 (7078): 803. https://doi.org/10.1038/439803a
- Poland, Therese M., and Deborah G. McCullough. 2006. "Emerald Ash Borer: Invasion of the Urban Forest and the Threat to North America's Ash Resource." *Journal of Forestry* 104, no. 3: 118–24. https://doi.org/10.1093/jof/104.3.118
- Popovic, Srdja. 2015. *Blueprint for Revolution: How to Use Rice Pudding, Lego*

*Men, and Other Nonviolent Techniques to Galvanize Communities, Overthrow Dictators, or Simply Change the World*. New York: Spiegel & Grau.

- Prunckun, Hank. 2019. *Counterintelligence Theory and Practice*. Lanham, MD: Rowman & Littlefield.
- Rabinovich, Abraham. 2004. *The Yom Kippur War: The Epic Encounter That Transformed the Middle East*. New York: Schocken.
- Rakove, Jack N. 2010. *Revolutionaries: A New History of the Invention of America*. New York: Houghton Mifflin Harcourt.
- Randolph, Marc. 2019. *That Will Never Work: The Birth of Netflix and the Amazing Life of an Idea*. New York: Little, Brown.
- Rasanayagam, Angelo. 2005. *Afghanistan: A Modern History*. London: I.B. Tauris.
- Rockwell, Theodore. 1992. *The Rickover Effect: How One Man Made a Difference*. Annapolis, MD: Naval Institute Press.
- Rose, Alexander. 2007. *Washington's Spies: The Story of America's First Spy Ring*. New York: Bantam Books.
- Rotary International. 2020. *PolioPlus: Rotary's Role in Eradication*. Evanston, IL: Rotary International.
- Rumsby, Ben. 2021. "'A Legislative Bomb': How Boris Johnson Helped Blow Up the European Super League." *The Telegraph*, April 21. http://www.telegraph.co.uk/football/2021/04/21/legislative-bomb-boris-johnson-helped-blow-european-super-league/
- Salk, Jonas. 1972. *Man Unfolding*. New York: Harper & Row.
- Sanger, David E. 2012. *Confront and Conceal: Obama's Secret Wars and Surprising Use of American Power*. New York: Crown.
- SANS Institute. 2023. *Common Insider Threat Patterns*.
- Sasson, Steven. 2013. "A Brief History of the Digital Camera." *IEEE Spectrum*, June.
- Schelling, Thomas C. 1980. *The Strategy of Conflict*. Cambridge, MA: Harvard University Press.
- Scheuer, Michael. 2004. *Imperial Hubris: Why the West Is Losing the War on Terror*. Washington, DC: Potomac Books.
- Schorger, A. W. 1955. *The Passenger Pigeon: Its Natural History and Extinction*. Madison: University of Wisconsin Press.
- Schrittwieser, Julian, et al. 2020. "Mastering Atari, Go, Chess and Shogi by Planning with a Learned Model." *Nature* 588: 604–609. https://doi.org/10.1038/s41586-020-03051-4
- Securities and Exchange Commission. 2013. "In the Matter of Knight Capital Americas LLC, Respondent." Administrative Proceeding No. 3-15570, October 16.

- Seneca the Younger. 1917. *Moral Letters to Lucilius*, Letter 13. Translated by R. M. Gummere. Cambridge, MA: Harvard University Press.
- Shakespeare, William. 2003. *Julius Caesar*. Edited by David Daniell. London: The Arden Shakespeare.
- Sharp, Gene. 1973. *The Politics of Nonviolent Action*. Boston: Porter Sargent.
- Shell Group Planning. 1983. *Scenarios 1973–1983*. London: Royal Dutch/Shell Group.
- Sheridan, Greg. 2015. "Ronda Rousey Fight Results and Statistics." ESPN. http://www.espn.com/mma/fighter/_/id/2489290/ronda-rousey
- Shine, Richard. 2018. *Cane Toad Wars*. Berkeley: University of California Press.
- Silver, David, et al. 2018. "A General Reinforcement Learning Algorithm That Masters Chess, Shogi, and Go through Self-Play." *Science* 362: 1140–1144. https://doi.org/10.1126/science.aar6404
- Silver, David, Aja Huang, Chris J. Maddison, et al. 2016. "Mastering the Game of Go with Deep Neural Networks and Tree Search." *Nature* 529: 484–489. https://doi.org/10.1038/nature16961
- Silver, David, Julian Schrittwieser, et al. 2017. "Mastering the Game of Go without Human Knowledge." *Nature* 550: 354–359. https://doi.org/10.1038/nature24270
- Sinek, Simon. 2019. *The Infinite Game*. New York: Portfolio/Penguin.
- Smith, Douglas K., and Robert C. Alexander. 1988. *Fumbling the Future: How Xerox Invented, Then Ignored, the First Personal Computer*. New York: William Morrow.
- Smith, Peter. 2014. "The Falklands War Logistics Timeline." *Journal of Military History* 78, no. 2: 589–612. https://doi.org/10.1353/jmh.2014.0096
- Southeastern Conference. 2006. *2006 SEC Championship Game Book*. Atlanta: Southeastern Conference, December.
- SpaceX. 2021. "Starship Testing Overview." SpaceX, March. http://www.spacex.com/vehicles/starship/
- *Sports Illustrated*. 1990. "Tyson Shocker: The Night the Myth Exploded." *Sports Illustrated*, February 19.
- *Sports Illustrated*. 1996. "Faldo Calmly Cruises as Norman Crumbles." *Sports Illustrated*, April 22.
- *Sports Illustrated*. 2015. "Tyson vs. Douglas: 25th Anniversary." *Sports Illustrated*, February 10. http://www.si.com/boxing/2015/02/10/mike-tyson-buster-douglas-25th-anniversary
- Springettsbury Township Board of Supervisors. 2019. Meeting minutes, March 28.

- Sternberg, Robert J. 1988. *The Triarchic Mind: A New Theory of Human Intelligence*. New York: Viking.
- Steuben, Friedrich Wilhelm von. 1779. *Regulations for the Order and Discipline of the Troops of the United States*. Philadelphia: Styner and Cist.
- Stone, Brad. 2013. *The Everything Store: Jeff Bezos and the Age of Amazon*. New York: Little, Brown.
- Sun Tzu. 1963. *The Art of War*. Translated by Samuel B. Griffith. Oxford: Oxford University Press.
- Surowiecki, James. 2010. "Last Blues for Blockbuster." *The New Yorker*, October 18.
- Symantec (Nicolas Falliere, Liam O. Murchu, and Eric Chien). 2011. *W32.Stuxnet Dossier*. Mountain View, CA: Symantec Security Response.
- Symantec (Geoff McDonald, Liam O. Murchu, Stephen Doherty, and Eric Chien). 2013. *Stuxnet 0.5: The Missing Link*. Mountain View, CA: Symantec Security Response.
- Taleb, Nassim Nicholas. 2007. *The Black Swan: The Impact of the Highly Improbable*. New York: Random House.
- Team Rubicon. n.d. "About Us." http://teamrubiconusa.org
- Tesla, Inc. v. Tripp, No. 3:18-cv-00293-MMD-WGC (D. Nev. December 10, 2020).
- Thompson, Kimberly M., and Radboud J. Duintjer Tebbens. 2014. "Lessons from the Polio Endgame: Overcoming the Final Hurdles." *Journal of Infectious Diseases* 210 (S1): S475–S483. https://doi.org/10.1093/infdis/jiu447
- Tromp, John. 2016. "Number of Legal Positions in Go." John Tromp's Homepage. Last modified January. http://tromp.github.io/go/legal.html
- Tuchman, Barbara W. 1962. *The Guns of August*. New York: Macmillan.
- Tyson, Mike, and Larry Sloman. 2013. *Undisputed Truth*. New York: Blue Rider Press.
- Ultimate Fighting Championship. 2015. *UFC 193: Ronda Rousey vs. Holly Holm* [Video]. YouTube, uploaded by UFC, November 15. Video, 3:17:30. http://www.youtube.com/watch?v=3j7WaTd8bkQ
- United Nations. 1945. "Agreement for the Prosecution and Punishment of the Major War Criminals of the European Axis, and Charter of the International Military Tribunal (London Agreement), 8 August 1945." London: United Nations.
- United States Army. n.d. "Birth of the U.S. Army (June 14, 1775)." Washington, DC: Office of the Chief of Military History.
- U.S. Army Ordnance Branch. 1962. *Small Arms Ammunition Study: Caliber*

*Transition Analysis*. Aberdeen Proving Ground, MD: U.S. Army Ordnance Branch.

- U.S. Bureau of Economic Analysis. 2020. *Gross Domestic Product, 2nd Quarter 2020 (Second Estimate) and Corporate Profits, 2nd Quarter 2020 (Preliminary Estimate)*. Washington, DC: BEA.
- U.S. Census Bureau. n.d. "Retail Video Store Revenues, 1990–2010."
- U.S. Department of Agriculture (USDA). 2021. *Emerald Ash Borer Program Manual*. Animal and Plant Health Inspection Service.
- U.S. Department of Energy and Department of the Navy. 2019. *Occupational Radiation Exposure from U.S. Naval Nuclear Propulsion Plants and Their Support Facilities*. Washington, DC: Naval Reactors/DOE.
- U.S. Department of the Treasury. n.d. "Treasurer of the United States." Washington, DC.
- U.S. Drug Enforcement Administration. 2003. *DEA History Book: A Tradition of Excellence 1973–2003*. Washington, DC: U.S. Drug Enforcement Administration.
- U.S. House Committee on Oversight and Reform. 2022. *Hearing on the Capital One Data Breach*. May 13.
- U.S. Securities and Exchange Commission. 1999–2011. *Kodak Form 10-K Filings*. Washington, DC: SEC.
- U.S. Securities and Exchange Commission (SEC). 2018. "SEC Charges Theranos, CEO Elizabeth Holmes, and Former President Ramesh 'Sunny' Balwani with Massive Fraud." Press Release 2018-41, March 14. http://www.sec.gov/newsroom/press-releases/2018-41
- U.S. Senate Select Committee on Intelligence. 1994. *Statement on the Aldrich Ames Espionage Case*. March.
- United States v. Microsoft Corp., 253 F.3d 34 (D.C. Cir. 2001).
- United States Holocaust Memorial Museum (USHMM). 2020. "International Military Tribunal at Nuremberg." Washington, DC: USHMM.
- USNI News. 2018. "USS John Warner Launches Tomahawks in Syria Strikes." April 14. http://news.usni.org/2018/04/14/uss-john-warner-launches-tomahawks-in-syria-strikes
- van Creveld, Martin. 1985. *Command in War*. Cambridge, MA: Harvard University Press.
- Van der Kolk, Bessel A. 2014. *The Body Keeps the Score: Brain, Mind, and Body in the Healing of Trauma*. New York: Penguin Books.
- Vaughan, Diane. 1996. *The Challenger Launch Decision: Risky Technology, Culture, and Deviance at NASA*. Chicago: University of Chicago Press.

- Von Neumann, John, and Oskar Morgenstern. 1944. *Theory of Games and Economic Behavior*. Princeton, NJ: Princeton University Press.
- Wack, Pierre. 1985. "Scenarios: Uncharted Waters Ahead." *Harvard Business Review* 63, no. 5: 73–89.
- Wainwright, C. Martin. 2004. "Entente Cordiale in Cartoon and Caricature." *History Today* 54, no. 4: 42–48.
- Wainwright, C. Martin. 2017. *Crispus Attucks York 2016 Annual Report*. York, PA: CAY.
- *The Wall Street Journal*. 2012. "A Tale of Human Error at Knight." *The Wall Street Journal*, August 2.
- Wang, Tony Tong, et al. 2023. "Adversarial Policies Beat Superhuman Go AIs." In *Proceedings of the 40th International Conference on Machine Learning (ICML 2023)*. arXiv:2211.00241. https://arxiv.org/abs/2211.00241
- Watson, Richard. 1995. "Netscape's Wild Ride." *BusinessWeek*, December 11.
- Weber, Thomas. 1997. *On the Salt March: The Historiography of Gandhi's March to Dandi*. New Delhi: Oxford University Press.
- Weiner, Tim, David Johnston, and Neil A. Lewis. 1995. *Betrayal: The Story of Aldrich Ames, an American Spy*. New York: Random House.
- Whaley, Barton. 2007. *Stratagem: Deception and Surprise in War*. Norwalk, CT: EastBridge.
- Whitby, Andrew. 2020. "If You Want to Go Fast ..." Blog post, December 25. http://andrewwhitby.com/2020/12/25/if-you-want-to-go-fast/
- White House Office of the Press Secretary. 2003. "President Bush Announces Major Combat Operations in Iraq Have Ended." May 1.
- *The White House*. 2020. *The President's Coronavirus Guidelines for America: 15 Days to Slow the Spread*. Washington, DC: The White House, March 16.
- Wilson, Jonathan. 2021. "The Core Problem of the Super League: A Lack of Soul." *Sports Illustrated*, April 19. http://www.si.com/soccer/2021/04/19/european-super-league-soul-tradition-history-greed-liverpool
- Winkeljohn, Mike. 2016. Seminar notes, Jackson-Wink Academy.
- Womack, James P., Daniel T. Jones, and Daniel Roos. 1990. *The Machine That Changed the World*. New York: Free Press.
- Wood, Gordon S. 2002. *The American Revolution: A History*. New York: Modern Library.
- World Heart Federation. 2024. *CVD Prevention*. Geneva: WHF.
- World Health Organization. 1988. *World Health Assembly Resolution WHA 41.28: "Global Eradication of Poliomyelitis by the Year 2000."* Geneva: WHO.
- World Health Organization. 2021. *Polio Eradication Strategy 2022–2026: Delivering on a Promise*. Geneva: WHO.

- World Health Organization. 2023. *Field Guidance for the Implementation of Environmental Surveillance for Poliovirus*. Geneva: WHO.
- World Health Organization. 2025. "Statement of the Forty-Second Meeting of the Polio IHR Emergency Committee under the International Health Regulations (2005)." Geneva: WHO.
- World Health Organization, South-East Asia Regional Office. 2014. "WHO South-East Asia Region Officially Certified Polio-Free." New Delhi: WHO SEARO.
- Wright, Robert K. Jr. 1983. *The Continental Army*. Washington, DC: Center of Military History, United States Army.
- Yergin, Daniel. 1991. *The Prize: The Epic Quest for Oil, Money, and Power*. New York: Simon & Schuster.
- Zetter, Kim. 2014. *Countdown to Zero Day: Stuxnet and the Launch of the World's First Digital Weapon*. New York: Crown.
- Ziegler, Martyn, and Matt Lawton. 2021. "European Super League: All Six Premier League Clubs Begin Withdrawal Process." *The Times*, April 20. http://www.thetimes.co.uk/article/european-super-league-all-six-premier-league-clubs-begin-withdrawal-process-f6b0t0qf9

www.ingramcontent.com/pod-product-compliance
Lightning Source LLC
Chambersburg PA
CBHW050643270326
41927CB00012B/2853